The United Nations – South Africa

*George J. de Lint*

# THE UNITED NATIONS

## The abhorrent misapplication of the Charter in respect of

# SOUTH AFRICA

W. E. J. TJEENK WILLINK B.V.
*Publishers*
ZWOLLE-HOLLAND

ISBN 90 271 1151 0

# Contents

*Appendices*

# Foreword

by
Professor Wijnand Langeraar, Delft, University of Technology
Rear Admiral Royal Netherlands Navy, ret.

It is with a feeling of pride and gratitude that I have accepted to introduce this publication to the widest possible international forum. All those who have cherished and fostered the high ideals which led to the creation of the United Nations must regard the present developments with a keen sense of apprehension and foreboding.

It seems as if every debate to review the relationship between South Africa and the United Nations develops into an indiscriminate cannonade aimed at South Africa's expulsion from that Organization. Rather than to try and take an objective look at the present situation against the perspective of promised developments and in comparison to what was, the subjective argumentation used by certain United Nations' members gives rise to the impression that their mind is made up and that they do not wish to be confused by the facts.

The United Nations, indeed, is now facing its hour of truth. Are we going to accept, for whatever political reasoning, the expulsion from the Organization of countries whose system of government does not conform to the views and beliefs of others? In that case a lot of weeding has to be done, which may well leave behind it a barren tract of land.

Can we accept the tortuous reasoning that South Africa now is a member of the United Nations because of the veto of three permanent members of the Security Council? Is that what the Charter provides for? Is that in keeping with its spirit?

Or does the Charter provide for the application of processes of conciliation, peaceful means for the solution of problems which are recommended to its members? Conciliation implies, of course, the willingness to understand the standpoint of the other party, based on, and supported by, objective statements of fact.

This publication is a serious attempt to provide facts objectively and unemotionally. It is one of the few documents published doing so and as such may constitute a major contribution to a future solution. It does not start from preconceived standpoints, but brings back to our memory what has been stated, has been decided, has been laid down. It is an honest document deserving to be read, digested and acted upon.

Let us not delude ourselves. Unless we look squarely at the situation in which the United Nations finds itself today, there will come a time when we bitterly deplore our present attitude of 'laissez faire, laissez aller'. Let us beware, lest our conceptions of right and wrong are eroded, gradually destroyed.

Let us take heed of the signs on the wall or our name one day will also be expunged from the membership list. Therefore, as Benjamin Franklin once said, let us hang together, lest we will be hung separately.

February 1976
WIJNAND LANGERAAR

# I

# South Africa, a short introduction

Holland's involvement in South Africa's affairs and in the post-World War international controversy surrounding that country stems primarily from the close bond existing between Dutchmen and South Africans. It was the Dutch, who in the seventeenth century, founded the settlement at the Cape and who first established themselves there. Later they were joined by settlers of French and German origin, and these were the people who, after the occupation of their colony by the British in the early nineteenth century, hived off northwards and eastwards in organised groups to found republics of their own – only to have these fall prey in their turn to British imperialism. The century-long struggle of this people the Boers for freedom waged against British domination was eventually brought to an end with the creation of the Union of South Africa, whereby the Afrikaners, as they were now called, at long last gained the right of self-determination – albeit with a certain restriction: under the new dispensation they had to share their self-determination with the British colonists who had settled in the colony in large numbers in the course of the nineteenth century.

People of the Netherlands throughout the centuries have felt closely involved in the struggle for freedom. The individual Dutchman is aware of this bond when visiting South Africa and it is felt collectively by the Dutch when South Africa finds itself under criticism. This is a perfectly legitimate involvement and no more to be decried as 'racialism' than is the attitude of Indians towards those of their kindred who, having settled in African countries, have been expelled from those countries or discriminated against.

Hollanders are far better equipped to understand the relationships and the general situation in South Africa because of their grasp of the Boer

(Afrikaner) language and also because their way of life, their culture and many of their social institutions, are familiar to them. Visitors from Holland cannot fail to be impressed by the honesty and sincerity of purpose with which South Africans are wrestling with their very considerable problems. Moreover, they discover that in many ways mental attitudes are very similar. Why, they think to themselves, these people are not strangers; they are very much like us! We Dutchmen are in a position to appreciate the South Africans' difficulties and in an important sense—given our historical, cultural and emotional links with that country—we have an obligation to encourage a better understanding for its successive governments in the present international environment of hostility towards South Africa. The problem which South Africa faces in how to work out an equitable system of rights for all of its diverse peoples. It is not a unique problem. But South Africa experiences it on a scale which is unique.

To secure for each of the country's major population groups the assurance that it will not be swamped by the others, that it has a future in which it will be able to realize its legitimate aspirations—this is the goal which the South African government has set itself. And this is something which the Dutch in particular should help the world to understand. We should help the world to understand that South Africa is composed of different embryonic fatherlands or national homelands which reality guarantees to each of the major population groups the opportunity to work out its own future. This applies not only to the various Bantu-speaking peoples, but in equal measure to the people of Dutch and other European descent.

However, the visitor to South Africa is struck by the fact that this symbiosis of peoples and fatherlands, so widely divergent in their characteristics and aspirations and in their stages of technological development, entails enormous problems of which the solution is still in full process. Thus, while the territorial core of the Homelands has been demarcated and one (the Transkei) will become independent during 1976, the consolidation of the Bantu homelands has not yet been finalised, and finality has also not been reached on their inter-relationship and each one's place in the general scheme of things. South Africa, in other words, is in a state of transition. A new order is in process of being born, a new basis being laid down for the inter-relationship of the various peoples to replace the pre-World War II patriarchial order. Given the diversity of the issues involved and the scale of the problems, this is clearly something which cannot be achieved overnight.

There is widespread agreement among South Africans on the need for change. While there certainly are differences in approach, the fact is that there is a strong feeling that the historically determined minority structure of the South African governmental system and discriminatory race relations practices must be changed—and these indeed are being changed. In this respect developments in South Africa reflect increasing pressure of change, generated both within and from without the society. Some of the more obvious elements of this pressure are the increased political influences of persons of colour and the increased effect, within South Africa's internal policies, of external factors notably as a result of events elsewhere in Southern Africa.

As far as race relations are concerned, South Africa has indeed moved out of the old era of white dictation into a new era which is characterised by dialogue and negotiation. This is strikingly illustrated by the deliberations which are occurring to an increasing extent between the central government leaders and the political leaders of all population groups. The meetings held in 1975, between the Prime Minister and responsible cabinet ministers on the one side, and the political leaders of the Coloured, Asian and Black communities on the other, are an example. This is a development of great importance and promise for the future, because it means that outstanding points of friction and conflict will increasingly be settled by reference to inter-élite negotiation.

A person studying the internal political situation in South Africa has to recognize all these factors and judge them on their merits. Certainly, he will not find it easy to gain a true image, the more so because the political situation in that country is coloured by the after-effects of the old colonial system and the old struggle between Boer and Brit together with their animosities in regard to the language medium. In the years since Union (1910) these two sections developed a close-unit community with equal sense of purpose. More and more South Africans—regardless of language—have come to realize that the ultimate success of the evolution to selfdetermination is no less than a matter of survival.

The question of survival has become predominant for the South African Government in its domestic and foreign policy as well as in relation to the measure in which concessions can be made in response to demands and pressures to change social structures within its borders. Dutch observers facing this reality have come to respect the attitude of the South African Government.

In the Netherlands, the identity of the South African nation, which has waged a successful struggle for its independence, and which Great Britain deemed worthy of investing with the responsibility of sovereignty, is not merely a vague concept, but an historically demonstrable and irrefutable fact.

The view expressed by King William III of the Netherlands on the occasion when the British Empire for the first time relinquished its territorial claims in the face of the South Africans' determination to create a state of their own, seems to us to be particularly telling. This happened when Britain recognised the independence of the Orange Free State in terms of the Bloemfontein Convention of 23 February, 1854. It was thereby agreed that the inhabitants of the Orange Free State were thenceforth to be regarded as an independent nation, completely free in every respect, and therefore no longer subject to the British Crown—a subjection which had been unilaterally imposed on them in 1848. The thereupon consolidated Orange Free State Republic was invested with a flag and a coat of arms by the King of the Netherlands. This Charter *sui generis* was elucidated by King William III in the following terms:

> 'Through circumstances beyond the control of my ancestors and yours, the happy people and the flourishing country now governed by you, were deprived of their authority. This deprivation was deeply deplored by us, because the sincere affection which all Holland felt for these people, was equalled by that of the latter for the House of Orange and the Netherlands. Even as in the olden days my august ancestor, William the Silent, and his successors, with their loyal Netherlanders, fought in the sacred cause of liberty, enlightment and civilisation, so you, a courageous outpost, have been waging an unremitting battle against the forces of barbarism and injustice which are constantly threatening to overwhelm you.
>
> May your endeavours be crowned with lasting success and may your struggle proceed with the least possible prejudice to your welfare.
>
> May this shoot from the original Dutch tree, which has rooted itself in your country, there to grow and develop its own characteristics, and to give rise to a new Dutch nation, continue to flourish; and may your fame and prosperity which, as I have noted with gratification and appreciation, you have linked to the name of my royal House, increase and multiply.'
>
> *(Translation)*

King William III's view of the South Africans and their constituting a courageous outpost in the face of great and often unknown dangers, which were constantly threatening to overwhelm them, was not only strikingly true more than a century ago, but has lost none of its validity. The same view was expressed repeatedly and explicitly during the years around 1954 by Dr.

Verwoerd, the Dutchborn South African statesman. The South Africans were fated throughout the 19th century to contend with pillaging, cattle raids and violence practised by neighbouring peoples, who constituted a perpetual menace to property and life. Then, as though all this were not enough, they still had to contend with what King William III had so aptly termed injustice. While on the one hand the neighbouring peoples were continually violating agreements and treaties in regard to rights and frontiers, the British authority on the other all too frequently proved an unjust arbiter, failing to redress the wrongs suffered by the Boers at the hands of their neighbours, and thereby adding the burden of injustice to the depredations.

These twin burdens—depredations on the one hand and injustice on the other—were temporarily lightened when, during the early years of the 20th century, the South Africans managed to wrest definitively from the British the right of self-determination. But after World War II the same two threats once more raised their heads, albeit in a new shape. Now South Africa faced externally-manipulated terrorism on the one hand and on the other injustice and unfair treatment at the UN which is aimed at supporting terrorism and weakening South Africa's right to defend itself. Struggling against these pestilences, the Afrikaner people is fighting for its survival.

As early as 1946, political tempests were unleashed against South Africa at the UN, and these have continued unabated without any sign of a change down to the time of writing. Such a change could conceivably be brought about by a firmer stand on the part of those Members who do not regard this development as conducive to the equitable application of the UN Charter and to the authority of the international body in its capacity as the guardian of the principles of justice and international law.

This should be the task of those Western powers who, by virtue of their historic ties with South Africa, are in a position to be far better informed about, and have a much better grasp of conditions in that country, and of the pros and cons of the social and political order and its evolution, and of the basis and aims of government policy, and of the implementation of the principle of self-determination, than the average Member of the UN.

The Netherlands in particular should not find it difficult to give other UN Members some idea of the manner in which the South African descendants of the Dutch, who founded the original settlement at the Cape, subsequently struggled successfully for self-determination, first in the Boer Republics, and later in the four Britishruled colonies. It would be a simple matter to describe unambiguously the identity of the nation that has accomplished this feat.

Naturally this nation has absorbed many alien elements, and especially its interassimilation with the population of British descent, has influenced its identity to a marked degree. But under the South Africa Act, passed by the British Parliament in 1909 (a matter which will be discussed in greater detail further on), the further evolution of this process of assimilation became exclusively the concern of the South African people themselves, being an aspect of their right of self-determination.

If one accepts the fact that in this year of 1976 it is simply no longer feasible to deprive a people of its right of self-determination, it should also be clear that this cannot be attempted either directly or indirectly, i.e. by way of enforced modification of the identity of such a people by dissolving it with other peoples in order to create a new entity, which will then be invested with the right of self-determination. Nobody has the right—and this includes the UN—to forcibly fuse a number of utterly heterogeneous peoples with different styles of living and at different stages of technological development; and emphatically not retrospectively, where one or more of these peoples have already attained independence.

The people, referred to in the Dutch Monarch's letter, managed to constitute itself as an independent Afrikaans 'Nation', governing the flourishing African country as it was described and appreciated by the King as long as 120 years ago. The King's letter presented a factual acknowledgment of the right of self-determination of the People particularly in relation to the sacred cause of liberty, enlightment and civilisation. This was the basis of the development of the people and the United Nations will never be able to demand that such development should in 1976 or later be undone. If ever this should happen, the cause of liberty, enlightment and civilisation would no longer be safe in the hands of the United Nations.

The Afrikaner people, also received its due recognition from authoritive British quarters when its identity and position as a separate African people was acknowledged so well by Mr. Harold Macmillan, the former British Prime Minister in his address to the South African Parliament in February 1960. Immediately after referring to the 'winds of change' which were blowing throughout Africa Mr. Macmillan said: 'Of course you understand this better than anyone. You are sprung from Europe, the home of nationalism. And here in Africa, you have yourselves created a free nation, a new nation.

Indeed, in the history of our times, yours will be recorded as the first of the African nationalisms!'.

Mr. Macmillan's statement may have spurred the South African Government to maintain the new free nation but also to respect the African nationalism of the other peoples within its borders who have not yet reached the threshold to freedom and independence. The South African Government heeded this prompting in accordance with the UN's ideals and it went to great sacrifice to pave the road to freedom and independence for those other peoples still in a stage of development.

This was, as the following chapters try to show, entirely in harmony with the Charter, and the ideological desire to thwart this development can only have arisen amongst those who wished the whole of South Africa's wealth to fall in hands of those underdeveloped peoples at the sacrifice of the Afrikaner people's own identity and achievements. As a 'white' people the afrikaners would no longer be allowed a place under the African sun. Could racial spite be the basis of this ideological desire?

The UN's crusade against the social conditions in South Africa commonly referred to as the 'policies of apartheid' degenerated into a misapplication of the Charter from the very moment these actions were subjected to the ideological desire to deny the Afrikaner people its own identity and to rob it of its achievements and inheritance. From that moment the UN have been operating under the banner of 'anti apartheid', a fals banner, and this shift in attitude has taken place so gradually and has been so well camouflaged by emotions and sentiments, that it has not been proved sufficiently possible for the democratic powers adequately to counter this misapplication of the Charter. As a consequence South Africa has been manoeuvred into a most awkward position and the question arises whether the democratic powers will as yet be able to develop a stategy which can save South Africa from its situation of international isolation.

Great Britain in particular has a special responsibility towards the future of this country. If in the following pages mention is made of this special responsibility, this is done without any pretension to throw doubt upon the wisdom of the British diplomacy. On the contrary, the author believes that the United Kingdom on the international level has the capacity to avert a catastrophic development in respect of South Africa. As a permanent member of the Security Council it has already given proof of its willingness and its ability to do this in 1974, 1975, and notably during January 1976 (see page 75).

II

# How the South African Afrikaner People achieved self-determination

The legal position of the Republic of South Africa as a Member of the UN, invested with absolute sovereignty over the territory under its jurisdiction, is unassailable; and moreover, South Africa has the indisputable right to demand that with the UN and other international organisations its sovereignty should rank no lower than that of other states. It is absurd for her detractors to challenge this sovereignty, with which the South African Government was invested long ago after the effectuation of the right of self-determination. It is therefore an unalterable fact that this Government does wield sovereignty within its own area of jurisdiction.

South Africa is already a decolonised country which has had its sovereignty bequeathed to it by Great Britain in 1931. The South Africans' struggle for selfdetermination was waged intermittently from 1806, when the Cape was conquered by the British, until 1926, when in terms of the Balfour Declaration the British Dominions were proclaimed 'autonomous communities within the British Empire, equal in status, in no way subordinate one to another in any aspect of their domestic or external affairs, though united by a common allegiance to the Crown, and freely associated as members of the British Commonwealth of Nations'. Then, five years after the Balfour Declaration, i.e. in 1931, the Statute of Westminster put the seal on the termination of each and every form of formal subordination of the governments of the Dominions to the British Crown, whereby the absolute sovereignty of South Africa, Canada and the other Dominions was legally established.

During the period 1835-1838, large organized groups of South Africans emigrated from the Cape Colony, to establish new republics in territories beyond British jurisdiction. They founded the South African Republic (the Transvaal), whose independence was recognised by Britain in 1852; and in 1854, in terms of the Bloemfontein Convention, Great Britain also accepted and recognized the complete independence of the Orange Free State, whose inhabitants, released from their subjection to the British Crown, were thereby also enable to found their own republic.

The history of the South African people's progress towards independence centres mainly around the two republics' struggle for freedom.

In 1876 England annexed the Transvaal; five years later, in 1881, the First Boer War resulted in a Boer victory, and the restoration of the independence of the Transvaal.

In 1899 Britain launched the Anglo-Boer War, a bloody and long drawn-out struggle which ended in the capitulation of the Boers; so that under the Treaty of Vereeniging of 1902, both republics lost their independence and became British Colonies.

During the period 1902-1909, a cold war was waged for the restoration of independence, in which the Transvaal and the Orange Free State were supported by the Cape and Natal. This cold war ended in the South Africans' regaining the effectuation of their right of self-determination through the creation of the Union of South Africa, whose Constitution was the South Africa Act, drafted by the National Convention and enacted by the British Parliament. The South Africa Act guaranteed self-determination, and, among other matters, provided for the recognition of the Dutch language, which the English during their one hundred and twenty years of domination, had done their best to suppress, as one of the official languages.

The South Africa Act, which was signed by King Edward VII on 20 September 1909 and came into effect on 31 May 1910, formed the foundation of the unified state i.e. the Union of South Africa, composed of the four British colonies. Although this Act went a long way towards granting self-determination to the South Africans, they were nevertheless still subject to certain restrictions, arising from the Union's legal position in relation to Great Britain. However, the decision in 1914 to declare war on Germany, had not been forced upon South Africa by Great Britain; it was a voluntary step, constituting an act of self-determination. Britain nevertheless retained a partial and limited formal responsibility for the evolution of South Africa' internal legal order until 1931, when South Africa and the other Dominions were granred complete independence under the Statute of Westminster. The final chapter in the long struggle for total independence was written in 1961 when South Africa became a Republic and relinquished its ties with the British Commonwealth.

The struggle for self-determination culminated in the unification of the four colonies. This political unification of 1909 was preceded by an intercolonial conference in 1908 (the Customs Conference), where unification was approved in principle. It was further agreed that the parliaments of the respective colonies would send delegates to a National Convention which was to be called for the purpose of preparing a draft constitution.

The National Convention first met in Durban on 12 October, 1908, was resumed in Cape Town and Bloemfontein, and wound up its work on 11 May, 1909.

The result was that a balance was struck between the political aspirations of the Boers and of the South Africans of British descent; moreover, a political system was evolved in terms of which guarantees were given for the restoration of the Dutch language and the cultural values which the Boers had treasured in the days of their independence before their subjection to the British Crown.

Many of the provisions of this Constitution were derived from the original constitutions of the South African Republic Transvaal and the Republic of the Orange Free State.

The administrative system of the former republics had related exclusively to the White community, between whom and the surrounding non-White peoples the most rigid segregation had been maintained. The Government service, as well as the electorate, was therefore composed exclusively of people of European stock. Their policy was one of non-interference in the affairs of the neighbouring Bantu peoples, at least in so far as such peaceful co-existence was feasible. This separation characterised relations between Whites and Bantu at the time of the National Convention, and the latter groups at that stage did not aspire to representation at the Convention for the purpose of discussing matters touching them.

The principle that the administration of the state was to be exclusively in the hands of the White population, i.e. the group represented at the National Convention, was inherent in the policy of the unified state. This policy was accepted as self-evident, albeit with certain minor restrictions, relating to the franchise in the Cape Province and Natal. These restrictions, however, did not prejudice this basic principle in any way; rather did they serve to consolidate it, since it was understood that at no stage in the future were they to be extended.

This was the decision of the National Convention, and this was how it was incorporated in the South Africa Act as drafted by the National Convention, which was signed on 11 May, 1909 by all delegates to the Conference and subsequently approved in the Transvaal, the Orange River Colony and the Cape Colony by the respective parliaments, and in Natal by referendum.

Then, finally, this decision was accepted also by the British Parliament in passing the South Africa Act, the preamble of which refers to the 'terms and conditions they have agreed to by resolution of their respective Parliaments'. This concise statement was expressed by the attitude of the British Parliament, namely its recognition of the four South African colonies' right of self-determination in all matters on which the National Convention had agreed.

At the National Convention in 1909, the problems in connection with the relations between Dutch-speaking and English-speaking South Africans caused the respective leaders of these two groups a good deal of brain-racking. They tried hard to find a common denominator for the two groups' aspirations towards self-determination.

The Boers, with the War of Independence still fresh in their memories, insisted on full independence; the English South Africans, on the other hand, were quite happy to remain completely subservient to British rule, and were eager to consolidate the existing ties with the motherland. That a fusion was possible, was due to the fact that the two groups had so much in common, culturally as well as in their views on the legal and social order; another factor was the circumstance that the two groups were on much the same level of development. In view of all this, Britain eventually agreed to withdraw and to grant self-determination to the newly created state, as demanded by the Boers. Considering the conflicting views of the two groups, it was inevitable that the resultant granting of independence was, in many respects, in the nature of a compromise. And so it has remained. To this very day there are certain elements in both population groups who have throughout and consistently refused to accept this compromise. Small factions of malcontents continued to agitate, and have gone on doing so right up to the present time. This accounts for the fact that important English newspapers in South Africa have never ceased opposing many aspects of developments which, incidentally, would possibly have followed exactly the same pattern even if Britain had not, in accordance with the National Convention of 1909, relinquished its jurisdiction over South Africa and South Africa's affairs.

The attacks from outside on the South African legal order, which commenced systematically after World War II and have continued ever since, are based almost entirely on a total disregard of the historical fact that as far as the Boers, as well as South Africans of British extraction are concerned, their demand for the right of self-determination was finally and most definitely acceded to by Great Britain with the passing of the South Africa Act in 1909.

In making use of its right of self-determination a people may, if it so desires, increase its numbers by admitting groups or individuals of other peoples, or even by assimilating with other peoples, always provided that these other peoples wish to be so assimilated. But where no such desire for assimilation or fusion exists on the part of the peoples concerned, any attempt by outsiders, whether the UN or another state, to enforce integration, is in direct conflict with the right of self-determination.

What is more, such enforced assimilation of an alien element cannot possibly be imposed upon a people if the foreseeable result would be that such an integrated people would, as far as its social order, legal institutions, culture, economy, etc. are concerned, exercise its right of self-determination by making all kinds of decisions which the orginal peoples so different in character, aspirations and level of development, could never subscribe to.

It is this consideration which should be taken into account by those who are so ready to criticise and counsel the South African government. This government cannot and should not be expected to agree to the fusion into a single integrated unit of the population of Western European descent with the numerically far superior Bantu-speaking peoples, when the aspirations of all these various groups, their views on justice, their culture, their economy, etc., are poles apart.

The only thing which the South African government can do to invalidate such criticism, is to grant to those Bantu peoples the right of self-determination within their own territories. It would be a grave mistake to reject this solution out of hand by labelling it as a 'policy of apartheid'. The autonomy idea is inherent in the principle of self-determination, and it cannot be disputed that this very principle is clearly recognised and firmly entrenched in the UN Charter itself.

III

# Great Britain and the policy of multi-national development

With regard to South Africa's policy of multi-national development, which has been so mercilessly criticised by so many outsiders, Great Britain is more of an insider, because the foundations of that policy evolved during that period when South Africa was still part of the British Empire; and even after South Africa had attained independence, the existing system was upheld and developed in accordance with the general principles of the legal order which had been created earlier under the British Administration. One might say that, until 1931, Great Britain was in fact co-responsible for the *status quo*.

The relations and phenomena referred to above, entailed the co-existence of a number of utterly heterogeneous peoples under the British flag. Everyone of these peoples had its own history, culture, aspirations and identity, all of which everyone wished to preserve intact. Naturally this frequently led to clashes with each other. To resolve these differences, settlements were made occasionally in the form of treaties and even at that time these settlements provided for bilateral separation. Now, although all this happened long ago, and even though the settlements made at that time have long since expired chiefly as a result of British annexation, they have not lost their significance, inasmuch as they serve to disprove the proposition, which is constantly and repeatedly stated at the UN, that all inhabitants of the territories formerly joined together under the British flag are now to be regarded as 'a people as a whole'.

The joining together of all these heterogeneous peoples under British rule was never aimed at interassimilation, and certainly not at intermarriage and integration. Wherever these peoples encountered one another, social forms of symbiosis and trade structures developed. There was exchange of capital, of labour and of other economic values. As time went on, the most important of these became the influx of Bantu workers into enterprises owned by people of European origin. In view of the difference in development (measured in technological terms) between the two groups, this inevitable resulted in the emergence of inequalities of opportunity. The exchange of capital and labour naturally led to a migration from the Bantu territories to those inhabited by the white South Africans first to the farms, and afterwards to the mines and industries. This migration increased very gradually and almost imperceptibly, yet it was this factor that in due course led to a formalization of the segregation, in other words, to a variety of measures designed to ensure that this physical invasion would not result in assimilation and racial friction between the groups. By modern standards many of these measures are unacceptable and should be abolished as soon as possible e.g. the laws concerning segregation in trains, buses, parks, etc. It should be remembered, however, that when many years ago these measures were instituted they were intended solely to safeguard the culture, identity and prosperity, in other words the general values of the European-oriented population, since it was feared that these values might be threatened by migrants whose culture and general standards were altogether different. It was not so much a question of race and colour; these were merely incidental factors. What considerably mattered were the totally different identity and culture of these migrant peoples. These measures should therefore not be regarded as evidence of calculated racialism and racial discrimination or be condemned as such.

The mistake the British Administration made was that it failed to take any corrective stepts to control the mass migration of Bantu labourers and their families from their own territories into other parts of the country, and failed to realise the effect this would have on social relations. The result has been that throughout the country, outside the Bantu-owned territories, Bantu townships have mushroomed, especially around the big cities and the industrial centres. This development naturally created all kinds of problems, which the authorities had to solve somehow. In this way the British Administration bequeathed to its successor a somewhat chaotic legal order; and now this successor is faced with the problem of doing away with segregation, which has become irreconcilable with the modern world's ideas of justice. Therefore Great Britain can only be expected to be extremely circumspect when participating in the annual debate on race relations in South Africa at the UN. In fact, it behoves her to be more than just non-committal, as she has a duty to defend South Africa to explain how this segregation came about in the first place, being herself very largely responsible.

But what has actually happened?

In elucidating these matters, it would also be necessary to explain what the premature abolition of obsolete measures as referred to above would involve and what the effect of this would be on the security of the country. The laws requiring people to carry identity papers, passes, or whatever one cares to call them, are essential in a country like South Africa, to which access is so easy. In every other country in the world where comparable conditions prevail, similar security measures have been found necessary. In South Africa's case these requirements are decried by her critics as 'apartheid measures'; but even if this were true, violation of these laws still does not entitle the transgressor to a halo. The mere fact of breaking the law does not turn a man into a 'freedom fighter', qualifying for international support and even encouragement for defying the laws of a country, just because that country happens to be South Africa, whereas in any other part of the world such law-breakers are dealt with as they deserve.

It is an absurdity to justify the actions of persons who perpetrate bomb outrages and other acts of terrorism in South Africa as an expression of their opposition to official race relations policy. Yet, absurd though it be, this attitude has gained so much sympathy in other countries, that it has become the 'done' thing to shield such people from the South African government 'on humanitarian grounds'—even in cases where the crimes in question were committed against black authorities, who derive their authority from the achievement of self-determination by the Bantu-speaking peoples.

Another matter which requires to be viewed in its true light is the measure aimed at redefining the territories of the various population groups within the framework of the policy of separate development. In principle these measures have nothing to do with (petty) apartheid and even less with discrimination; to determine their fairness or unfairness, one should see them in their true perspective: they can only be considered and criticized in their relation to the policy of multi-national development and not in abstracto.

Evaluation of the various measures referred to above requires insight and a knowledge of conditions in the country, of the history, the legal order and the aspirations of the various peoples with respect to the evolution of that legal order; and those critics of South Africa, who have done so much to turn world opinion against that country, are either ignorant of the facts or refuse to acknowledge them. It has appeared useless for the South African government to attempt, during UN debates, to argue against such stubbornness and lack of understanding. Moreover, these debates concern matters falling within the domestic jurisdiction of South Africa, and which in terms of Article 2 (7) of the UN Charter exclude intervention on the part of the General Assembly, not only formally but also practically, inasmuch as other Members, lacking the knowledge and insight referred to above, are quite incompetent to pass judgement. But, for historic reasons South Africa might look to Great Britain for support in UN deliberations, and to expect her to oppose decisions which place South Africa in the wrong.

However, not only has Great Britain done little of the kind, but after a number of years of reticence, she suddenly and without warning turned against South Africa; and from then on has actually supported decisions which the British Government, being better informed, should have opposed, especially in view of the fact that Britain herself shared responsibility for the creation of that very situation in South Africa which is now being condemned in UN resolutions.

Up to 1961 Great Britain categorically and unreservedly opposed any interference by the UN in South Africa's domestic affairs. Thus British representative, Lord Birdwood, addressing the General Assembly on 9 November 1959, presented his country's point of view as follows:

> If . . . on this occasion we once more register the vote against the resolution as a whole, it is because we still feel that it contravenes the charter and that the need to maintain the integrity of the charter is of paramount importance and must take priority over other considerations. Article 2 (7) was placed conspicuously in its governing position at the beginning of the charter after long and careful consideration. 'Nothing contained in the present charter', it states—that is nothing that follows thereafter, for example, Article 10 or Articles 55 and 56—shall authorise the United Nations to intervene in matters which are essentially within the jurisdiction of any state.*

In 1961, however, Great Britain made a complete *volte face*, as is evidenced by the statement of her representative Mr. B. Smithers, during a meeting of the Special Political Commission on 5 April.

Mr. Smithers expressed himself as follows:

* Report on the proceedings of the Fourteenth Session of the General Assembly of the United Nations held at New York, September 15 December 13, 1959. CMND. 992. Annexe. 7, page 122.

> Apartheid confronts us with circumstances which are, so far as I know, unique in the annals of this Organisation. We see the deliberate adoption, retention and development of a policy specifically based on total racial discrimination. This is further distinguished by the circumstance that it is discrimination amongst and against the permanent inhabitants of the country itself. Such a policy, which is a deliberate exaltation of descriminatory principles, stands alone in its category.
>
> Furthermore, it clearly has grave international repercussions, particularly in Africa, but also in other continents. The outcome of the Commonwealth Prime Ministers' Conference recently held in London makes this quite clear in a practical and unmistakable fashion.
>
> While, therefore, the importance we attach to the proper observance of Article 2 (7) of the Charter remains undiminished, we regard the case of apartheid, in the circumstances which now exist, as of such an extraordinary and exceptional nature as to warrant our regarding it and treating it as *sui generis*.*

One is prepared to accept that when Great Britain suddenly adopted this new attitude towards South Africa, she did not do so with any perfidious purpose; that it was not her intention suddenly to betray the South African Government, but rather under influence of the Commonwealth Prime Ministers' Conference to create an opening for herself to offer to act as mediator—a mediator who would by virtue of this change of front be more acceptable to those UN Members who by that time were firmly convinced of the 'grave international repercussions', as well as to South Africa, who, it was expected, would be prepared to turn a blind eye to Great Britain's somewhat sudden change of front.

Great Britain could not be expected to imagine that Article 2 (7) of the Charter should apply to apartheid in every other country with the exception of South Africa. Neither could she, in the course of a mere eighteen months, have so changed her views on the nature of South Africa's internal policy that in 1961 she could aver that Article 2 (7) was not applicable in the case of South Africa, whereas in 1959 she had asserted just the opposite. If Britain had really experienced a change of heart, the *volte face* should have been in the opposite direction, in view of the developments in South Africa, where it was becoming increasingly obvious that the implementation of the policy of multi-national development was actually helping to create a climate conducive to the abolition of obsolete situations and the creation of a new legal and

* Report on the proceedings of the Fifteenth Session of the General Assembly of the United Nations held at New York, CMND. 1456 (United Nations (No. 1) 1961), page 144: Speech on the question of Race Conflict in South Africa by Mr. P. Smithers, M.P., in the Special Committee on April 5, 1961.

political order in which discrimination would no longer feature at all.

If discriminative situations are to be eliminated from the legal order which was originally created by Great Britain in South Africa, two prerequisites must be observed: Firstly, its abolition must be accomplished in such a way as to be compatible with the various peoples' right of self-determination; and secondly, and even more importantly each people's right of self-determination must be duly safeguarded. Great Britain has done little to further the development of the Bantu peoples' progress towards self-determination, little more, in fact, than laying the necessary foundations as incorporated, for example, in the Native Land Act, (Act No. 27 of 1913). These foundations having been duly laid, the South African Government, as Britain's legal successor, could not absorb the administrative structures of the Bantu peoples within their own territories into the unitary structure already existing in other territories within the Republic, or into a single structure yet to be devised, and designed for the whole country with all its divergent peoples. What might be feasible is an umbrella structure in the form of a confederation or something along those lines; but such a system on a bilateral or multilateral basis will be workable only when the black nations have attained self-determination, as this will be the only basis on which they will be able to participate in full-scale decision-making.

As soon as black self-determination has become an accomplished fact, the necessary conditions will exist for the abolition of all forms of obsolete apartheid. This will in all probability be achieved on a basis of bilateral or multilateral agreements in which due provision will be made for the legal position of individuals of one people within the legal territory of another, e.g. with regard to human rights, land ownership, social provisions and political rights. This will also guarantee majority rule in each and every one of these self-governing territories and if not majority rule, whatever form of government a people for itself wishes to adopt.

The principle of partitioning the Bantu-speaking peoples and the people of European descent in their respective territories was already firmly established long before the Union in 1909; and it continued to apply during and after the Union, as is apparent from the South Africa Act, which provided for the separate administration of the Bantu territories i.e. the territories which belonged to the Bantu traditionally and historically. In fact, the Union Parliament regarded this matter as a separate problem, distinct from those created by the unification of the four British colonies, and one requiring co-ordinating legislation. (Cf. Prof. C.F.J. Muller: *500 jaar Suid-Afrikaanse Geskiedenis*, p. 345.).

And at an even earlier stage, the British administration had tackled this problem by appointing a commission under the chairmanship of Sir Godfrey Lagden, which during the period 1903-1905 submitted a number of recommendations; these were incorporated in the Natives Land Act of 1913. In terms of this Act, territories covering 10,4 million morgen in all were reserved for Bantu ownership. The same Act provided that a commission be appointed to determine whether this reservation required to be modified. Following the recommendations of this Commission, which was headed by Sir William Beaumont, the Bantu territories as established in 1913 were subsequently (i.e. in 1936) considerably extended.

In the years that followed, this policy was continued, and had the active co-operation of the British authority. When the General Assembly of the UN in 1971 declared the establishment of Bantu homelands in South Africa to be a reprehensible step constituting 'a violation of the inalienable rights of the African people and contrary to the principle of self-determination' (G.A. Resolution 2775E (XXVI) of 29 November 1971), this was a challenge to the British delegation, which ought to have taken it upon itself to inform the Assembly of the truth of the matter by pointing to the very considerable progress already made by those very homelands towards self-determination.

When the homeland Transkei in 1976 will have acquired its independence and subsequently will apply for membership of the United Nations, Great Britain, again, will have a special responsibility to support the application. In November 1975 at Umtata, the capital of Transkei, this special responsibility was referred to in a speech by the Chief Minister of Transkei, Mr. K.D. Matanzima, following a pronouncement in UN circles that Transkei would not then be admitted. So very much to the point Mr. Matanzima said:

'In no less a measure than virtually every Black nation in Africa, were Transkeians the victims of European imperialism and colonialism of the past centuries. It is purely an accident of history—over which we had no control whatsoever—that our colonial rulers, the British, transferred their sovereignty over us to another power, the Union of South Africa.

Had this not happened the power from which the Transkei's imminent independence would come would have been the United Kingdom and not the Republic of South Africa. In this event, if one is to be guided by the examples of Botswana, Lesotho, Swaziland, Zambia, Malawi, Uganda, Kenya and Tanzania, I presume we would have been welcomed into the UN, as were they. But because of an accident of our birth we are to be denied memberschip.'

It passes my comprehension, therefore, why the anti-colonial UN clings to those imperial relics so passionately, and rejects the Transkei which is doing nothing than regain sovereignty over its traditional territory.'

IV

# The South African Bantu Peoples before and on the march to self-determination under British rule and later

Furthermore, the British delegation was in a position to advance valid arguments correcting other misrepresentations contained in the abovementioned Resolution, e.g. the passage 'noting' that the Government of South Africa, 'while treating the White inhabitants of that country, irrespective of their national origins, as constituting one nation, seeks artificially to divide the African people into "nations" according to their tribal origins and justifies the establishment of non-contiguous Bantu homelands on that basis.'

In this regard it was only necessary to point out that the groups occupying the various Bantu territories are in effect distinct peoples, inasmuch as each one possesses its own language, culture, traditions, political aspirations and preferences in the sense of the UN Charter, which distinguish it from other peoples. These different peoples and their present situation are listed below:

| *Territory* | *Population* | *People* | *Extent of territory* (ha) |
|---|---|---|---|
| Zulu | 4.026.082 | Kwa Zulu | 3.271.667 |
| Xhosa (Transkei Ciskei) | 3.929.000 | Transkei | 3.855.692 |
| | | Ciskei | 939.626 |
| Tswana | 1.718.508 | Bophutta Tswana | 3.806.274 |
| North Sotho | 1.603.530 | Lebowa | 2.221.114 |
| South Sotho | 1.453.354 | Basotho Qwagwa | 48.243 |
| Shangaan | 736.978 | Gazankulu | 633.110 |
| Swazi | 498.704 | Swazi | 206.253 |
| Ndebele | 416.641 | South Ndebele | 20.243 |
| Venda | 357.875 | Vendaland | 619.956 |

This information clearly shows that there is no question of insignificant little tribes or of a conglomeration of African tribes who have been 'artificially' segregated for the purpose of dispersing and weakening the Bantu, but that these are in fact peoples of considerable numerical importance and that if, having attained independence, they should qualify for membership of the UN, their position among the Member nations would be such as to entitle them to play an equal role in that body. Naturally these various peoples may, at some later stage, and if they so desire, effect a fusion; but judging by, for example, the present attitudes of the Transkei and the Ciskei, such a development is unlikely. These two territories are contiguous, the language is the same, and tribal origin so closely related that one would expect them to move towards inter-assimilation. But this is by no means the likely case. They have made it abundantly clear that they are opposed to such fusion; and there is no reason to suppose that the other Bantu people would take any more kindly to suggestions of this kind. A possibility is that some of these peoples would seek fusion with Lesotho, Swaziland or Botswana, which have already attained membership of the UN.

In the South Africa Act no clear provision was made for the self-determination of the Bantu peoples. The Act did, however, provide for control over and separate administration of Bantu affairs. This was the responsibility of the British Government, and by succession, under the new legal order, of the Governor General in Council. The relevant sections of the South Africa Act further provided for a separate legal status for the Bantu peoples within their own territories.

These provisions, based as they were on conditions prevailing before 1909, allowed the Bantu peoples to pursue their own way of life in their own homelands. As far as system of government, social institutions, cultural customs and economy were concerned, interference by or on behalf of the Governor General was restricted to the bare minimum. These peoples therefore enjoyed a considerable measure of autonomy; but that they would eventually aspire to self-determination was not foreseen at that stage and no provision was made for this. The emergence of these aspirations after World War II confronted the South African government with a problem which it proceeded to solve in a manner that was in no way contrary to the UN Charter. The government not only recognised and sympathised with these aspirations, but took active steps towards their realisation, spending vast amounts of money on development aid to further the economy and to provide social, educational and other services.

There is no reason why this solution of the problem of granting self-determination to the Bantu peoples should not carry the approval of the UN, especially in view of the fact that Great Britain followed much the same policy in regard to the three Protectorates which, though originally included in South Africa, remained under British jurisdiction. Section 151 of the South Africa Act provided that the administration of these territories, 'inhabited wholly or in part by natives', could be delegated by the King to the Governor General in Council. Great Britain, however, did not act in accordance with this provision. Instead, these territories were granted full independence in the mid-sixties. This is how Lesotho, Botswana and Swaziland were created. This was done with the wholehearted approval of the UN; yet this same body rejects the solution envisaged by the South African government in regard to the Bantu territories in the Republic, which, under the South Africa Act, from the very beginning came under the jurisdiction of the Governor General in Council. It must therefore be concluded that the UN is given to applying double standards. The injustice of this attitude becomes quite glaring when one considers that Great Britain had spent relatively little on development aid in the three territories in preparing them for independence. In other words, she was not prepared to make sacrifices comparable to those being made by South Africa in its honest efforts to create the necessary opportunities for the peoples presently governed by her to progress towards true independence.

The peoples of the other national homelands within the confines of South Africa are just as qualified as the people of Lesotho, Botswana and Swaziland to take their own future in their hands and to pursue their own political ambitions and preferences. In fact they have already gone a long way in that pursuit. They have come to possess of their own legislative councils and conforming governmental structures and they have themselves produced statesmen and politicians to lead them. These leaders are the ones whose actions and utterances express these ambitions and preferences for all the world to note.

So it can be seen that these peoples are in the process of realising their goal of self-determination with a sense of purpose and dedication rarely met in other parts of Africa. They tolerate opposition in politics and respect the freedom of expression. When these peoples responded to the South African Government's policy of separate development, that policy was accepted. It is important to note that such acceptance was not given in subservience or in ignorance, but with all sorts of 'nuances' having regard, among other things, to plans later to be conceived for further co-operation. In fact Mr. Matanzima, Head of the State of Transkei together with Mr. Buthulezi of KwaZulu launched plans for a federation in which also the neighbouring States of Swaziland and Mozambique are to be incorporated and which in territory would exceed that of the Republic of South Africa.

A people's right of self-determination does not extend beyond its own territory. This applies to white South Africans as much as to the various Bantu peoples. Those critics who complain that the homelands belonging to the black nations are too small to support their respective populations, ignore the fact that the creation of the homelands was not a matter of arbitrarily demarcating certain areas and assigning them to specific Bantu peoples. It is a matter of fact that these homelands are the very territories originally occupied by the peoples concerned—territories within which they had long enjoyed a measure of independence, enough, at any rate, to enable them to follow their own way of life and uphold their own social order and administrative structures, as suited to their particular identity and aspirations. The consolidation of the Bantu homelands has been nothing more, or less, than the creation of the necessary opportunities to enable these peoples to progress towards self-determination and to exercise their rights in a manner compatible with modern ideas and with rapid modern developments. Thanks to the practical aid provided by the South African Government in the educational field and in boosting the economies of these homelands, this development is proceeding apace, at a rate which few other African states can hope even to equal.

These Bantu territories and their boundaries have existed for a very long time; and if these territories are in any way inadequate, or their boundaries not clearly enough defined, this is hardly the fault of the present Government, but a situation created over a period of more than a century by the British colonisers and bequeathed intact by them to the present authority. It is to the South African Government's credit that it is making great efforts, involving major sacrifices, to acquire more land to add to the Bantu homelands. The trouble is that some of these territories are scattered, and it is now the government's policy to consolidate these scattered entities. To achieve this, certain interchanges of land have been necessary.

The implementation of the relevant legislation will necessitate certain changes in the settlement pattern, to which the inhabitants of the affected areas (blacks and whites alike), may well object. There are situations however, where a balance is to be found between individual interests and the common good; and it should be realised that this legislation is aimed at hastening the independence of the peoples to which these individuals belong. That is why it is so absurd that of all measures particularly the reallocation of domiciles should be decried by the critics of the South African Government's policies as a 'violation of human rights and fundamental freedoms'.

The Bantu territories are far from being the arbitrary creations of the South African Government, their boundaries were defined long ago, by way of settlements, whereby each people accepted the other's territorial rights. It would, therefore, be incorrect to speak of 'Bantu reserves', because this term makes no allowance for the element of reciprocity, i.e. the recognition and acceptance by the inhabitants of these territories of the fact that the territories beyond their borders, though contiguous, do not belong to them but to other peoples, white or black. The territorial division of South Africa dates from the 19th century, when numbers of different peoples, who had come pouring into the country from other parts, encountered one another there. The Afrikaner people migrated northwards and eastwards from the Cape Colony, where there were no Bantu-speaking people at all. The Bantu peoples were mostly migrants from the north who had been driven southwards as a result of internecine wars. When the white trekkers arrived, they found a vast land which had either never been inhabited or had become depopulated through inter-tribal warfare and which the white immigrants from the Cape Colony were therefore free to occupy.

The 'wandering of the Nations' in the 18th and 19th century has been decisive for the pattern of their settlement in Southern Africa, as has been the case in Europe and elsewhere in the world in history. This pattern got its final shape when Great Britain was in authority.

Therefore the rights of the territory which they now inhabit cannot be considered to be inferior to those of other nations. It is not a case of the rights of individuals, settlers, wandering workers and others, but of those nations which after having wandered have found their final territory.

This is how the territorial division of South Africa came about; and under the British administration it was consolidated into the mosaic which Great Britain subsequently bequeathed to her successor, the Union of South Africa. This mosaic is by no means an ideal basis for the division into national homelands with the ultimate object of self-determination for the various peoples occupying them, but this is something for which Great Britain is partly to blame. For when this consolidation was being effected, the British administration allowed all sorts of seemingly petty little mutations to develop; this was the natural outcome of the colonial authority's shortsighted *laissez faire* policy, which took no heed of long-term developments.

It would, however, be unfair to lay all the blame for these developments on the British, since it is an indisputable fact that the territorial division had, by and large, been effected long before the turn of the century.

# V

# Self-determination and the ideal of realization of human rights

South Africa amply deserves recognition by other powers for its efforts to develop friendly relations with its neighbours and other African States on the basis of 'respect for the principle of equal rights and *self-determination of peoples'* (art. 1 par. 2 Charter). That is the main theme of the development according to the policy of 'separate development', which is being applied by South Africa within its domestic jurisdiction.

Thus South Africa is contributing to the realization of the second purpose of the United Nations as defined in article 1 (2) of the Charter. It seems clear that in doing so South Africa contributes at the same time to the realization of the—interrelated—third purpose of the United Nations (article 1 (3)), that is

> 'To achieve international co-operation in solving international problems of an economic, social, cultural, or humanitarian character, and in promoting and encouraging respect for human rights and for fundamental freedoms for all without distinction as to race, sex, language, or religion;'

Quite apart from the question whether the General Assembly of the United Nations would be compentent to dictate to South Africa a policy other than that of separate development, it should emphatically be pointed that this Assembly has never been able to suggest an alternative acceptable in terms of the principle of self-determination of peoples. For this very reason one must consider the objections voiced by the General Assembly year after year to be destructive and not to the point.

South Africa will be able to counter the objections to its official race relations policy by its recognition of the right of self-determination for the peoples within its borders; and the arguments of the opponents of this solution may be countered by pointing out that while this new order may in fact amount to a 'new-style apartheid', this is precisely the type of self-determination referred to and provided for in the Charter itself. The recognition of the right of self-determination is not aimed at discrimination but at the very opposite: it amounts in effect to the creation and safeguarding of values which will contribute to the equality of the fundamental and other rights of the individual members of communities which have attained a self-determined status.

As the ideal of self-determination for the various peoples is achieved, this will also become the basis for the solution of those problems—economic, social, cultural and humanitarian—which complicate interrelations between the various peoples living in South Africa, and lay the foundation for mutual co-operation and respect for human rights and fundamental freedoms throughout the Republic of South Africa. If the individual members of the various peoples are to enjoy full human rights and fundamental freedoms, the simultaneous step towards this ideal should be self-determination for each of these peoples as a whole. This vision is confirmed by the legal fact that the self-determination of peoples, which constitutes one of the principles of the UN Charter, has been incorporated in the 1966 Convenants on Human Rights as 'the *right* of self-determination for all peoples'. This 'human right' granted to the group will determine the quality of the 'human rights and fundamental freedoms' to be enjoyed by the individuals constituting that group; and it is this initial 'human right' which will prove to be of the most vital importance to these individuals. It follows, therefore that individual rights and the rights of the group as a whole are interdependent.

Many are the critics of South Africa who would deny this obvious fact, and who cannot or will not see that the question of individual freedoms is closely connected with the self-determination of the people as a whole. What these critics want is to overthrow the entire existing political structure in order to institute another structure by a one-man-one-vote system. But what they tend to forget is that this initial phase, i.e. the attainment of self-determination for each of South Africa's communities, is specifically provided for in the UN Charter itself and that the peoples of South Africa cannot be expected or required to forego it voluntarily. In fact, there is nothing to indicate that they are at all anxious to skip this phase. Another thing which these critics conveniently forget is that at least one of these groups, namely the people of mainly Western European descent, attained self-determination a long time ago and, since that time had become fully sovereign.

Those critics who want to deprive the established South African community of European descent of its sovereignty and right of self-determination and who want to hand the government of the country over to the other, numerically superior peoples, are outsiders, who rely mainly on the information supplied and the aspirations voiced by dissidents and malcontents, agitating mostly abroad. If these dissidents are to be believed, they are quite competent to take charge of everything—all assets and values—which the people of European descent have created and built up over many years; and they insist, moreover, that this take-over is to be effected according to the 'decolonisation' recipe applied in other African territories.

These critics should be given to understand that their recipe is not workable, for the simple reason that South Africa is already a decolonised country, inhabited by a number of different peoples, each of whom must progress towards self-determination on a basis of their own aspirations. South Africa is a country managed by a fully sovereign government invested with all the necessary power and authority to guide and direct its various peoples' progress towards ultimate independence.

The South African government has every right to object to the attacks and to all interference in its legal order and to resist those agitators who are primarily responsible for these attacks and who would not hesitate to resort to violence to gain their ends. This fundamental right actually constitutes the basis of the UN Charter, which declares that 'the Organization is based on the principle of the sovereign equality of all its Members (Article 2(1))', and is entrenched in the restriction on intervention by the Organization, which is precluded in principle from interfering in matters 'which are essentially within the domestic jurisdiction of any state' (Article 2(7)). The right and the function of a state to determine and evolve its own legal order are internal matters which fall within its sovereign jurisdisction and responsibility not covered by any special responsibility of the UN, and it is incumbent upon such a state to observe 'in good faith the obligations assumed in accordance with the Charter', in its internal no less than in its external dealings. This is what the Charter states, and this in turn implies a restriction of the Organization's authority to intervene. And should the UN consider it necessary at any time to question the internal legal order of any of its Members, it would be bound to respect that Member's own views and any *bona fide* solution suggested by the country concerned.

It is an accepted legal principle that a party's *bona fides* have to be accepted unless the contrary can be proved. In the General Assembly, where

South Africa's official policy of multi-national development is constantly under fire, the question of *bona* or *mala fides* tends to be overlooked. This is understandable, since those critics who keep hurling abuse at South Africa's official policy are unable to suggest any workable alternative. If these critics want to accuse South Africa of *mala fides* they ought to be able to come up with a fair and workable alternative. And this is where the democratic powers ought to raise their voices in protest against the vague and inadequately defined accusations which are hurled at South Africa.

South African policy decisions which are aimed at developing the country's internal legal order are inspired by regard for the right of self-determination of the various peoples dwelling within its confines; this approach is fully in line with the obligations as laid down in the Charter itself. In so far as the UN has any authority at all to evaluate these decisions, it would have to judge them on their merits in order to determine whether or not they are designed to further the progress of each of the peoples concerned towards self-determination.

In any case, this is a question of policy which is so obviously within the domestic sphere of South Africa that it is technically impossible for any outsider to evaluate them. The principle of the matter aside, the UN is in no position to do so since it lacks the necessary data on which such an evaluation could be based. The political judgements which the Organization has been passing upon South Africa were inspired by the political views and selfinterests of foreign powers and not by humanitarian considerations concerning the fate of the peoples within South Africa. They clash with the UN Charter whenever they infringe upon the process of the achievement of self-determination by those peoples. It is no less than folly when the UN would wish to block a people from attaining its independence. All recommendations from the General Assembly to its member-states not to maintain contact with the governments of what are they called Bantustans and not to recognize these in whatever way, are contrary to the UN Charter. Such recommendations are not only outside the competence of the General Assembly but also affect the prerogatives of the Security Council since they prejudice the future situation which will arise when these nations—after attaining independence—apply to the Security Council to be recommended for admission to membership in the United Nations (article 4 of the Charter).

There again it is up to the United Kingdom and the other democratic powers to throw a clearer light upon these questions and to refrain from making concessions in their speeches in the UN meetings which would in any way render the right of self-determination of the peoples of South Africa problematical.

VI

# The United Nations and the elimination of 'apartheid'

Since 1952 the United Nations have been conducting a strong campaign against the policies of 'apartheid' in South Africa, which is however pointless in so far as it fails to connect 'apartheid' with the reality of the separate development.

The point in question now is: is South Africa free to transform its legal order according to the policy of separate development, as discussed in previous chapters? Or would it be incumbent upon the United Nations to block such a development and to demand and enforce another, e.g. a development whereby control over the country is transferred by the Government to a 'black majority', whatever that may be?

One would say that no country would tolerate such interference in its own affairs which is quite contrary to the principle of sovereign equality of all UN Members as embodied in article 2 par. 1 of the Charter. One would do well to consider that whilst the United Nations may not obstruct separate development in South Africa, it logically follows that they cannot demand that South Africa abolishes or rescinds any measures which are considered essential by the competent authorities to achieve the goal of separate development, even if such measures are labelled by the United Nations as 'despicable apartheid'.

Of course anyone is free to condemn 'apartheid' in South Africa and to work for its abolition. One can in fact observe that the Government and public opinion in South Africa themselves are working for it and that what is here involved is a set of problems which can very well be solved through dialogue, negociation, conciliation or other peaceful means not only within South Africa but also beyond its borders. In that case however parties should carefully define the substance of what is called 'despicable apartheid'. Moreover, one should not analyse and condemn such 'apartheid' *in abstracto* but only *in concreto* i.e. in the context of the historically developed situation in South Africa and in relation to South Africa's own chosen way to modernise its legal order. To condemn 'apartheid' in its discriminatory aspect—in any country—only in abstracto is working with half truths. And this is precisely what has been done—to the detriment of South Africa—in the United Nations discussions and in the many reports called for on the subject by the United Nations.

When one criticizes the 'apartheid'-situation in South Africa in concreto i.e. in connection with the policy of separate development then one should be prepared to accept that the current apartheid-laws in this period of transition cannot be changed overnight. The transitory aspect of this position is predominant. It then also becomes clear that the perspective of the realization of that policy would imply the abolition of the apartheid situation.

The United Nations discussions and abovementioned reports of committees and experts have hardly bothered to criticize *in concreto*.

The outcome of the lengthy conflict between the United Nations and South Africa regarding the phenomenon of 'apartheid' in that country will depend on the following factors.

Firstly the time-element. The operation of the policy of separate development has progressed so much that in 1976 Transkei will become an independent nation; other states will follow rapidly thereafter. As a result situations of 'apartheid' which today may give rise to indignation will cease to exist and the subject of the conflict will automatically disappear.

South Africa may have to continue to resist the UN intervention until that time is reached, unless however the United Nations will be prepared to accept at an earlier date the truth of the new political structure emerging out of separate development. It is the reality of that new situation which in any case must mean the end of the conflict. It will prove to be stronger than the statement repeatedly made in the resolutions of the General Assembly, that the transformation of South Africa in separate nations should be discarded as a scheme only meant to keep control of the whole of South Africa out of the hands of the black majority. Whilst there are no indications that the General Assembly will be prepared to abandon its current views and to accept reality, neither is it to be expected that the General Assembly will be able to persuade the Security Council to allow it to implement its aspirations, to put back the clock and to undo what already has been achieved.

To be sure, it will not be easy for the General Assembly—once it is prepared to accept the reality of separate development—to break with the practices followed in respect of South Africa for so many years. For one thing, it would have to cease its officially proclaimed international campaign against apartheid, on which the following can be noted.

On 6 November 1962 the General Assembly resolved to establish a Special Committee to keep the racial policies of the Government of South Africa under review when the Assembly was not in session (Resolution 1761 (XVII)).

On 16 December 1966 the General Assembly by its Resolution 2202 (XXI) endorsed the proposals of this Special Committee for an international campaign against apartheid under auspices of the United Nations. In the same resolution the Assembly expressed its regret at the fact that the main trading partners of South Africa 'including three permanent members of the Security Council' had refused to join the Special Committee and by their increasing collaboration with the Government of South Africa had encouraged the latter to persist in its racial policies. In this resolution 2202 (XXI) the Secretary-General was asked to support the work of the Committee and assigned to the Committee were numerous activities, including resistance to tightening of economic and financial relations between other States and South Africa. Furthermore all States were recommended 'to discourage immediately the establishment of closer economic and financial relations with South Africa, particularly in investment and trade and also to discourage loans by banks in their countries to the Government of South Africa or South African companies'. Through these interlinked paragraphs of resolution 2202 (XXI) the Special Committee as well as the Secretary-General were mobilised not only against South Africa but also against 'the three permanent members of the Security Council'. The confrontation between the Assembly and these three permanent members—with their decisive votes in the Security Council—was beginning to show. As did the attack on the financial and economical position of South Africa and the notion that it would be sinful to invest in that country, a notion which in 1974 and later grew particularly in the circles of Churches and Labour Unions.

By Resolution 2307 (XXII) of 13 December 1967 the General Assembly requested and authorized the Special Committee to intensify its efforts to promote an international campaign against apartheid and to this end among other things in consultation with the Secretary-General to arrange for the advisory services of experts or special studies on specific aspects of the campaign. Moreover it asked the Committee to compile a report on measures which might appropriately be taken to ensure the widest dissemination of information on the evils of apartheid and the efforts of the international community to secure its elimination.

Thus the General Assembly started the international campaign against apartheid giving it a course different from the policy of the Security Council which—as will be discussed later—had decided in 1963 to achieve the elimination of apartheid in South Africa through peaceful means.

The campaign was meant to expose the 'evils of apartheid' and the Special Committee became a veritable factory of hostile propaganda against South Africa of the most virulent kind. The Committee never managed to define the substance of what could be considered evil apartheid, but merely labelled social relations in South Africa as 'apartheid' for no other reason than that these had to do with South Africa. It should be noted that no effort was made to prove the evil character of certain conditions in South Africa which elsewhere in the world also exist but are not considered evil. As an example may serve the legislation against terrorism (such as the Terrorism Act and the Suppression of Communism Act) which in the case of South Africa was termed unacceptable apartheid legislation, whilst similar legislation in other countries is considered reasonable and appropriate and never defined as 'apartheid legislation'.

By judging however the situation in South Africa *in abstracto* by ignoring the context of the historically developed general situation in South Africa as well as its impact on South Africa's own chosen way to modernise its legal order along the lines of separate development, the campaign worked with half truths. As a result its presentations became hostile propaganda which misinformed and infected world opinion.

For the purpose of exposing the horrors of the policies of apartheid, the General Assembly and the Special Committee have always presented this policy as one of oppression i.e. of such abhorrent oppression that the international community ought to come up in arms against it. However, the evidence of this oppression has almost solely been derived from the objection that in South Africa persons were brought to Court for having opposed 'apartheid'. On the basis of this sole objection, these persons used to be represented as 'the victims of repression' (see for 1975: A/10022, A/10281, A/SPC/174).

Such methods are dishonest and unworthy because the persons concerned were prosecuted and convicted not for their opposition to apartheid but for having committed murders and other crimes punishable everywhere in the world. This point will be reverted to in Chapter IX.

It is incorrect and misleading to accept the rendering of evidence of improper repression as an additional facet of oppression. The UN-reports and resolutions however consistently failed to advance more convincing proof. A student of these matters will find that whenever South Africa's legislation is labelled as a 'legislation of oppression' reference is made to the abovementioned Terrorism Act and the Suppression of Communism Act. It is true that these Acts authorize retention but only in respect of persons who nowhere in the world would be allowed to do as they please quite irrespective of whether they are for or against apartheid.

It is definitely not so that in South Africa persons are ever imprisoned or restricted in their freedom because they are just 'outspoken opponents of apartheid'. The subversive and terrorist activities which do form grounds for imprisonment are dealt with in Chapter IX. These acts rather that the South African defensive precautions ought to be labelled 'abhorrent' by the outside world.

To equate the policies of apartheid to a policy of oppression and repression is wrong and misrepresents the situation in South Africa. Apartheid or separation (seperate development) is not synonimous with oppression, but will indeed end a situation in which one people or group can oppress another.

The people of Transkei, and others aspiring after independence, can therefore appreciate it and this very fact brings to light an inconsistency in the policy of the General Assembly, which on the one hand would like to end oppression but on the other is blocking the independence of these peoples.

General Assembly Resolution 3068 (XXVIII) of 30 November 1973 misinformed and influenced world opinion out of all proportion. Through this resolution the text of an International Convention on the Suppression and Punishment of the Crime of Apartheid, was adopted, together with an appeal to all States to sign and ratify that convention as soon as possible. The convention in question could create havoc in the order of international law. Its contribution to the hostile propaganda against South Africa through the listing of a series of 'inhuman acts' which together would constitute 'the crime of apartheid' and which could generate 'similar policies and practices in South Africa', is false and most unfair for the following reasons.

It should be noted that the 'inhuman acts' mentioned in this list are punishable crimes in South Africa, as they are in many other states practising the rule of law. Crimes are mentioned like 'denial to a member or members of a racial group or groups; by murder of members of a racial group or groups' etc., and other serious crimes which are equally punishable in other countries of the world in which the rule of law applied. Only in the case of South Africa, however, are these serious crimes labelled as being 'inhuman acts' of 'apartheid'.

Since the inhuman acts mentioned in the convention are no part of the policies and practices of apartheid, the convention implies most notably that 'inhuman' acts which do belong thereto, could not be identified by the draftsmen the convention. Furthermore it is worthy of note that the inhuman acts—as was observed by Mr. Ferguson (USA) during the discussions in the General Assembly on 30 November 1973 (A/PV.2185)—had already been covered by the Genocide Convention and other existing conventions as a consequence of which the incorporation of these crimes in the International Convention on the Suppression and Punishment of the Crime of Apartheid, only served the political cause of identifying the South African authorities with these crimes.

In the Third Committee on 24 October 1973 (A/C.3/SR.2003) Mr. Wiggins (USA) said his Government had misgivings about the use of international conventions to achieve political ends, no matter how laudable those ends might be. In the General Assembly Mr. Ferguson (USA) (A/PV.2185) termed the convention unacceptable.

'Rules for the protection of human rights cannot be elaborated in such a manner as to ignore the very fabric of the rules of law. Efforts to protect human rights which ignore the rule of law can only lead to chaos'. The convention purported to effect a broad extension of international jurisdiction i.e. even in cases where there are no significant contacts between the offence and the forum State and where the offender is not a national of the forum State. All this was inconsistent with the basic norms of fairness, due process and notice so essential in criminal law.

Similar objections were raised in the same session of the General Assembly by Mr. Mackenzie (U.K.) (A/PV.2185): 'There are a number of features of the Convention which we find entirely unsatisfactory and unacceptable. One of the most important things is that it contains provisions which would violate the principles of international law concerning the proper exercise of criminal jurisdiction, principles to which we attach the highest importance. The provisions in question purport to autorize contracting States to exercise criminal jurisdiction in respect of certain matters covered by the Convention over acts done outside their jurisdiction by persons who are not their nationals. That assertion of jurisdiction would be totally inadmissible as far as my Government is concerned, and if this Convention should enter into force, my Government reserves its rights in relation to any attempt to assert jurisdiction over United Kingdom nationals. We believe that many Governments share our position'.

By accepting this convention the General Assembly gave the enemies of South Africa an instrument by which the relationships with South Africa could be rendered chaotic. If these enemies should claim jurisdiction over South African nationals on the strength of the convention, the South African Government would react. These actions and reactions could escalate into situations which would endanger international peace and security.

It is a fact that there are States keenly interested to exploit these unstable prospects for their own ends. During the discussions on the Convention in the Third Committee, Mr. Human (Yemen) raised the economic and political relations between South Africa and the Zionist regime, saying: 'Israel's policy has caused wars in the Middle East and threatened international peace and security and the policy of apartheid in South Africa, if not eliminated, would have the same results' (A/C.3/SR.2003). Thus Mr. Human applauded the Convention as an instrument to eliminate apartheid in South Africa, in a manner which suggested that the enemies of Israel might use that instrument, or a similar instrument to attack Israel by claiming jurisdiction over Israeli nationals. The chaotic effect of General Assembly resolution 3068 (XXVIII) by which the convention was adopted should definitely not be underestimated.

# VII

# The UN's approach before 1960

On 6 August 1957, Bolivia together with seven other states requested the Secretary-General to place 'the question of racial conflict arising from the South African Government's Apartheid policy' on the provisional agenda of the 12th General Assembly. In the course of the debates arising from this proposal, the South African representative objected to this item on sound legal grounds and received the unconditional support of Great Britain and Australia. But that this small minority would be outvoted was of course a foregone conclusion.

Admittedly, mention was made of the import of UN interference as laid down in the Charter. The concensus, however, appeared to be that in any event the Member states should be given every opportunity of blowing off steam and debating the matter as heatedly as they pleased, since the right of discussion in the General Assembly has never been subject to any restrictions. Decisions on the implementation of recommendations could be made on the basis of these discussions. As the Columbian representative put it, the UN had every right to investigate violations of human rights, but it could not intervene in isolated instances where a violation had been proven. The Iraqi representative, who was particularly critical of South Africa, observed that no Member-state had suggested that the UN take active steps to put an end to apartheid.

New Zealand and Canada announced their willingness to support a resolution reminding all UN Members of their obligations under of the Charter, but stopped short of lending their support to any action directed exclusively against South Africa.

Certain other representatives proposed that a solution be sought by way of mediation and conciliation. The Peruvian representative in particular, in seeking a solution, tried to take due account of the UN's functions and powers. He expressed the view that the UN should not drag the matter out indefinitely in the General Assembly for the sake of the UN's own honour and glory. To arrive at a solution, one would have to start again from the beginning: the necessary parties would have to be found to raise the question in the UN, which would then have to exercise its powers to attempt to solve the dispute by way of mediation and conciliation, impartially and without bias. The Peruvian spokesman went on to say that the UN should refrain from taking any action whatever, pending negotiations between the parties concerned, namely the Union of South Africa on the one hand and the champions of racial equality on the other. The parties would have to undertake not to cut short their negotiations, and should further be required to seek advice from the International Court of Justice in regard to the interpretation of the Charter and of existing treaties. If, after all this, the negotiations were to end in a deadlock, that would be the appropriate time for the General Assembly to take action.

It is a great pity that this sensible approach was disregarded by the Special Political Commission of the General Assembly.

On the other hand, however, proposals designed to pave the way for vexatious interference in South Africa's internal affairs, of which the UN had some experience over the past few years, also failed to gain support. An example of this was the proposal by Ceylon to the effect that the United Nations' Secretariat be required to submit annual 'progress reports' on the racial situation in South Africa.

It was only on the last day but one of the discussions by the Special Political Committee that a draft resolution was introduced by the Philippines and Ireland, to the effect that no measures should be instituted against South Africa; that the matter should not be brought up again in the General Assembly; and that the parties concerned should even refrain from initiating procedures for the purpose of pursuing the matter at all—even by way of mediation and conciliation.

The dogmatic exposition by the representative of the Philippines in regard to this draft resolution is of particular importance when it comes to determining the status which the South African issue should have enjoyed at the UN. The UN should have considered itself bound by this view, inasmuch as the majority of its Members were opposed to any intervention which would exceed the limits so explicitly formulated in this resolution.

Implicit in this official exposition was the fact that the UN operates at two levels: where questions of principle or doctrine are concerned, the Organization should resort to recommendations and moral persuasion; in the case of practical problems, to workable proposals; while in certain cases both methods will apply.

In regard to the matter at issue, the General Assembly, according to the Philippines should have operated at the former level. Consequently the architects of the draft resolution considered that no action was indicated beyond appealing to the South African government to uphold the lofty principles of the Charter, or to revise its policies in the light of the said principles and of world opinion. Moreover, South Africa was requested to inform the Secretary-General of its decision (Gen. Ass. Resolution 1178 (XII) of 26 November 1957).

It could be expected that South Africa would ignore this resolution; but this would not justify the UN to abandon this proposed approach, which was in accordance with the Charter, and to resort instead to enforcement measures not provided for in the Charter.

Whatever South Africa's reaction therefore, the UN had bound itself to adhere to the policy of recommendation and moral persuasion in promoting respect for human rights and fundamental freedoms, with no thought to resorting to violence. This was the gist of the resolution passed by the 12th General Assembly on 26 November 1957 (Resolution 1178 (XII)).

Moderate though this resolution appeared to be, Great Britain nevertheless considered it unacceptable—the British representative pointing out that Article 55 of the Charter, in which the directives to the Organization in regard to respect for, and observation of, human rights and fundamental freedoms are set forth, does not place Member states under any international obligation which could prejudice their domestic jurisdiction; and a country's policies governing her own citizens could surely not be regarded as anything but a domestic matter.

VIII

# The UN's approach since 1960

Until 1960 the UN's methods of intervention in South African affairs were still of such a nature that a solution by way of negotiation and conciliation seemed possible. At that stage the view also was that South Africa was to be given an opportunity of providing its own solution, and the UN was prepared to recognise the fact that South Africa was indeed making every effort to improve the relations between its various peoples and to bring its policies into line with modern thinking. It did not seem likely then that the more virulently anti-South African Members would succeed in persuading the UN to resort to non-peaceful forms of intervention in order to bring South Africa to its knees.

In 1960 the picture changed. In March of that year South Africa's enemies were provided with ample material for condemnation of that country's government as a result of a clash between the police and black demonstrators which led to the death of seventy persons. Immediately the whole world was up in arms. The Sharpeville incident was represented as tangible evidence of the oppression suffered by the Bantu and the direct outcome of the policy of apartheid. This was the only aspect of the matter which the propagandists were interested in presenting to the outside world. There were other aspects, too, but these were conveniently ignored, so that a fair and just evaluation of the Sharpeville incident—which shocked South Africans as much as it had the outside world—became impossible. This event provided the more virulently anti-South African Members of the UN, who had always been opposed to the idea of a peaceful solution, with a powerful weapon. Now, at last, they were able to sway the vote in the General Assembly and even to hope to induce the Security Council to exercise its powers under the

Charter in order to bring South Africa to heel. As a result of all this, from 1960 onwards possible UN intervention assumed a different character.

The stand taken by the Security Council encouraged South Africa's opponents to hope for that Organ's intervention in the form of economic sanctions and other violent means which the General Assembly had no authority to impose. On the other hand, the involvement of the Security Council meant that the matter was to be taken out of the hands of the General Assembly; in other words, that the latter body could not deal with the South African issue while it was being handled by the Security Council. This was in accordance with the clear directive contained in Article 12 paragraph 1 of the Charter, namely that the General Assembly shall not 'make any recommendation with regard to that dispute or situation' while the Security Council is exercising 'the functions assigned to it in the present Charter' in this regard. The General Assembly may, if it so desires, continue to discuss the question, but is precluded from taking any action whatever, and even from making recommendations or expressing any wishes in connection with the matter in hand.

But those Member states who had hoped to see the Security Council bring South Africa to its knees by means of enforcement action and had been consistently agitating for this in the General Assembly, were doomed to disappointment. The Security Council did not go to the lengths demanded by the majority in the General Assembly, either as regards the contents of the demands made of South Africa, or in regard to the measures to be adopted to enforce demands. The relations between the General Assembly and the Security Council consequently became obscure and strained indeed, all the more so when, after the Security Council had been induced to take the matter in hand, the General Assembly, in defiance of Article 12, continued to make recommendations, some of which were calculated to force the Security Council into taking more drastic steps against South Africa.

Notwithstanding the zeal of the General Assembly and the fierce determination of the majority of the members to subdue South Africa, the Security Council's intervention did not do it as much damage as might have been expected. In fact, it came out of the fray remarkably well, if not quite unscathed, despite certain side-issues, which will be discussed in greater detail further on.

In regard to the Security Council's intervention, there are a few points which require elucidation. The most important of these is the object of this intervention and the question of South Africa's jurisdiction; in other words, to what extent it should be allowed to regulate it own domestic affairs without interference on the part of the UN.

A comparison of the texts of the resolutions of the Security Council and those of the General Assembly, shows some glaring discrepancies. Both as regards the method of intervention and the nature of the demands made on South Africa, the Security Council and the General Assembly arrived at different conclusions. Prior to 1960 even the General Assembly had thought only in terms of 'a conciliatory approach in accordance with the principles of the Charter' as the only solution to the problem of inducing South Africa to modify her policies (G.A. Reslution 1016 (XI) of 30 January 1957). But after 1960 the tone and content of the G.A. resolutions ceased to be reconcilable with a 'conciliatory approach'. Thus the resolution of 28 November 1961 alleges that South Africa 'has failed to comply with the repeated requests and demands of the United Nations', and calls upon 'all States to consider taking such separate and collective action as is open to them' to bring about the abandonment of the reprehensible policies of the South African Government (G.A. Resolution 1598 (XV) of 28 November 1961).

In 1962 the General Assembly deviated even further from the policy of finding a solution by way of negotiation and conciliation, in that Member states were urged to apply specific enforcement measures against South Africa, e.g. breaking off diplomatic relations and boycotting South African products. Moreover, the Security Council was urged 'to take appropriate measures, including sanctions, to secure South Africa's compliance with the resolutions of the General Assembly and of the Security Council on this subject, and, if necessary, to consider action under Article 6 of the Charter', this being the Article providing for expulsion of a Member from the UN (G.A. Resolution 1761 (XVII) of 6 November 1962).

Nevertheless the Security Council could never be persuaded to adopt this vexatious policy, of which the recommendation by the General Assembly was, after all, hardly reconcilable with Article 12 of the Charter. In its Resolution 182 (1963) of 4 December 1963 the Security Council condemned 'the policy of apartheid' in no uncertain terms but added a rider to the effect that 'therefore a positive alternative to these policies must be found through peaceful means'. This pronouncement accorded with the policy of a 'conciliatory approach' which, prior to 1960, had been regarded even by the General Assembly as the only feasible one. The Security Council was not prepared to adopt measures against South Africa involving severance of diplomatic relations or economic boycotts, nor to encourage or support the relevant decisions made by the General Assembly in any way.

Lacking the Security Council's support, the General Assembly could not make much headway with its decisions in this regard. As stated explicitly in the Charter, the institution of measures such as 'interruption of economic relations and the severance of diplomatic relations' (Article 41 of the Charter) is emphatically the prerogative of the Security Council; and where the General Assembly feels that such measures are called for in cases being dealt with by the Security Council, all it can do is to await the Security Council's request (Article 12 of the Charter). For the General Assembly to make such recommendations to its Members direct when the Security Council is dealing with a matter is contrary to the Charter and therefore ultra vires.

Despite all the pressure exerted by the General Assembly to decide upon enforcement measures, the Security Council has hitherto refused to comply.

Another important point in the abovementioned Security Council Resolution is the fact that the Security Council has at least come up with concrete demands, which is more than the General Assembly has ever been able to do. This point is all the more important because the Security Council's demands are not irreconcilable with the legal order in South Africa and the way in which it is being developed under the aegis of the government, that is with due regard to the right of self-determination of the various peoples with their different identities, social structures and levels of development. The Security Council has sought the co-operation of the South African Government in order to bring about a 'peaceful and orderly transformation' of its obsolete social order, with guarantees of 'full, peaceful and orderly application of human rights and fundamental freedoms to all inhabitants of the territory as a whole' (Resolution 182 (1963) of 4 December 1963). This resolution does not deny the fact that the 'inhabitants' referred to belong to different peoples, the right of self-determination of each of which being an important factor in effecting the required transformation.

The General Assembly, however, has demanded something quite different, namely that South Africa should, without any transition, switch from her present system to one of 'majority rule based on universal suffrage'. According to the General Assembly, this was the only way in which 'the people of South Africa as a whole' could attain 'their right of self-determination' (G.A. Resolution 2396 (XXIII) of 6 December 1968 et seq.)

The Security Council, in terms of Resolution 182 (1963) of 4 December 1963, ruled that it would be necessary to find an alternative solution, one that could be realised by peaceful means. In this Resolution the Secretary-General was requested 'to establish a small group of recognised experts' to formulate such an alternative, 'under his direction and reporting to him'.

It is to be deplored that this 'group of experts' did not succeed in its task. The reason for this was that it was too deeply involved politically with hostile powers and organisations—hostile, that is, to South Africa—and was unable to bridge the gap sufficiently to enter into discussions with the South African Government. The group's finding that South Africa should be saddled with a 'National Convention' would not have been acceptable to the government of any sovereign state. Despite the impotence of the group of experts, the Security Council still held to its original course, i.e. of seeking a peaceful solution.

The group's report, submitted to the Council by the Secretary-General on 20 April 1964 (U.N. Doc. S/5658), did contain the recommendation that 'if no satisfactory reply were received from the South African government by the stipulated date on the proposal for a national convention, the Security Council would be left with no effective peaceful means for assisting to resolve the situation, except to apply economic sanctions'. However, the Council did not act on this recommendation.

The South African Government and the Group of Experts could not come to deliberation and understanding. South Africa could not receive the group. 'The visit of the Group, the South African reply stated, was not only specifically intended as interference in the internal affairs of the Republic, but the Group was also expected to prescribe how South Africa should be governed and, by implication, even what should be the provisions of its Constitution' (Yb. U.N. 1964, p. 108).

The Security Council rejected the proposal of the group of so-called experts that in the event of South Africa's refusing to agree to the proposed national convention, economic sanctions be applied to force her into compliance. Paragraph 5 of Resolution 191 (1964) of 18 June 1964, runs as follows: '. . . endorses and subscribes in particular to the main conclusion of the Group of Experts that "all the people of South Africa should be brought into consultation and should thus be enabled to decide the future of their country at the national level".' The idea of a National Convention was not adopted, nor was a specific procedure prescribed for the 'consultations among representatives of all elements of the population in South Africa'. (Paragraphs 6 and 7.)

In the course of the discussion, the United Kingdom representative had expressed the view that 'changes in South Africa must come about from within the country and should be acceptable to the peoples as a whole, in accordance with their rights and aspirations'. He went on to point out that any economic embargo required enforcement, 'and Members should not delude themselves that a peaceful solution and coercion could be reconciled' (Yearbook U.N. 1964, p. 113).

Whenever during the discussions in the UN meetings the formulation of demands on South Africa came up, rarely any attention was ever paid to the measures South Africa could take. We refer here to South Africa's maximum ability to adapt to foreign pressures to effect changes in its social structures without surrendering its sovereign identity. Usually the demands far exceeded the possible limits, which could be explained by the desire of the majority of the General Assembly's Members to finish off South Africa.

On these occasions the different viewpoints of the General Assembly and the Security Council became clear. The Security Council was never found prepared to go along with the General Assembly in posing demands upon South Africa aimed at causing its downfall.

This difference in viewpoints also became apparent when the Security Council refused to co-operate in the procedure to expel South Africa from the UN Organization to which we shall refer later on.

Only once in a UN meeting did anyone point out that it would never do for the UN—at least for the General Assembly—to dispose of a Member through the exercise of such pressures as could be foreseen to leave that Member no alternative than to act in self-defense. We would recall the address made during the discussions in the Special Political Committee on 9th October 1963 by the Danish representative Mr. Lannung.

Mr. Lannung in referring to the implications of the very form of transformation to impose upon South Africa which was to be advocated by the General Assembly, said it would be necessary to convince the white minority in South Africa that such a transformation would not necessarily mean the end of its existence. The whites would have to be reassured and prepared by a gradual process to accept a policy prescribed by the UN. 'One aspect of the problem' according to Mr. Lannung, was 'that under the Charter, practical steps of that nature were reserved for decision by the Security Council, and any Member State could disregard recommendations adopted by the Assembly alone. For that reason, Denmark considered that resolution 1971 (XVII)' (which recommended severance of diplomatic relations and a general economic boycott), 'while useful insofar as it proclaimed the principle of pressure, did not in itself bring the implementation of the principle much closer.'

Continuing, he said: 'It was impossible to predict with certainty what the results would be if the United Nations relied wholly on a policy of pressure, but there was a strong possibility that the results would be the opposite of what was desired. The European population in South Africa, left without hope of any acceptable alternative but evacuation, would entrench themselves still more in their misguided attitude, apply even more repressive measures against the non-white majority, and thus precipitate a major catastrophe. However heavily the responsibility for such developments would rest with the white minority, it must be the duty of the United Nations to do its utmost to prevent them'.

IX

# Terrorism and armed activities in and against South Africa

The vehement reactions throughout the world to the developments in South Africa since 1960 caused the General Assembly to concern itself with the incidence of terrorism in that country.

In South Africa terrorist attacks and schemes were reported, which had been organised and incited from abroad. Prior to 1965 the General Assembly had not clearly defined its stand in the matter of international terrorism. This was first done in the General Assembly Resolution 2131 (XX) of 21 December 1965, the text of wich appears as an appendix on page 88. It contains the 'Declaration on the Inadmissibility of Intervention in the Domestic Affairs of States and the Protection of their Independence and Sovereignty' hereafter referred to as 'Declaration on non-intervention 1965'. The Declaration on non-intervention 1965 provides that 'no State shall organize, assist, finance, incite or tolerate subversive, terrorist or armed activities towards the violent overthrow of the régime of another State or interfere in civil strife in another State.

From 1960 onwards, South Africa has been the victim of terrorist plots. Around 1960, the South African authorities discovered that the state was threatened by terrorist plotters who were being aided and abetted by persons and organisations abroad. The full extent of this conspiracy was disclosed when police raided the conspirators' headquarters in Rivonia, near Johannesburg, on 11 July 1963. A number of documents was confiscated, including detailed operational plans for terrorist activities of every conceivable kind. The data in connection with this case, known as 'Operation Mayibuye', were submitted to the UN by the South African Government. The guilty parties received a fair public trial in the South African Supreme Court. In the course of the trial, the authenticity of the documents was fully established, as was the guilt of the accused, who did not deny having committed the acts of terrorism with which they were charged. It was further established that Nelson Mandela and his accomplices had received support from Tanzania, Algeria and other African countries. The extent of their subversive activities may be deduced from the indictment prepared by the Attorney-General for the Province of the Transvaal. These activities included the recruitment of persons for military instruction and training, both within and outside the Republic of South Africa; the preparation, manufacture and use of explosives for the purpose of committing acts of violence and destruction in the aforesaid Republic; the art of warfare, including guerilla warfare and military training generally, for the purpose of causing a violent revolution in the aforesaid Republic; etc.

Shortly after 11 July 1963 (the date of the Rivonia raid), the UN took note of the relevant data and of the arrest of the parties concerned and their impending trial. The result was that the Security Council in its resolution 181 (1963) dated 7 August 1963 called upon the Government of South Africa to liberate all persons imprisoned, interned or subjected to other restrictions for having opposed the policy of apartheid. This call to liberate was clearly made in reference to the terrorists of the Rivonia plots although these were not specifically mentioned. France and the United Kingdom abstained from voting on this resolution.

In its resolution 1881 (XVIII) of 11 October 1963 the General Assembly specifically concerned itself with the South African court case against the Rivonia terrorists: 'Requests the Government of South Africa to abandon the arbitrary trial now in progress and forthwith to grant unconditional release to all political prisoners and to all persons interned or subjected to other restrictions for having opposed the policy of apartheid'. The request refers to 'the arbitrary trial' although it is common knowledge that the South African legal system precludes arbitrary trials, but, on the contrary, provides all safeguards to persons prosecuted for crimes, in accordance with the Universal Declaration of Human Rights. Everyone in South Africa is entitled in full equality to a fair and public hearing by an independent and impartial tribunal, in the determination of his rights and obligations and of any criminal charge against him. Everyone in South Africa charged with a penal offence has the right to be presumed innocent until proved guilty according to law in a public trial at which he will have all the guarantees necessary for his defence (articles 10 and 11 Universal Declaration of Human Rights). The court cases against the Rivonia terrorists were conducted in South Africa in strict accordance with those rules and under the careful scrutiny of many attending foreign experts.

In spite of the above the majority of the General Assembly decided to term the Rivonia trial 'arbitrary', thus outvoting the minority which could not accept the untruthfulness in the draft resolution. The manner in which the Charter was applied in this case was, to say the least, peculiar. The conclusion that the trial had been arbitrary could not be based on the statement incorporated elswhere in the resolution that the trial was arranged under arbitrary laws, since the criminal charges pressed did not relate to any arbitrary laws whatsoever.

The Security Council discussed the Rivonia trials in June 1964 and on 9 June 1964 it adopted Resolution 190 (1964) which also made reference to 'the arbitrary Rivonia trial' instituted against the leaders of the anti-apartheid movement and 'instituted within the framework of the arbitrary laws of apartheid'. Not only France and the United Kingdom, but also the USA and Brazil abstained from voting on this resolution. These four countries thus made it known that they had no desire to follow the majority on the road of arbitrary application of the Charter.

The abovementioned UN resolutions favouring the Rivonia terrorists are no longer topical. One would think that the General Assembly would have changed its attitude towards terrorism when it had gained the insights which became the basis for the Declaration on non-intervention 1965. There is good reason to doubt this however since the efforts on the part of the USA and other Western powers in 1972 to urge the UN into international action against terrorism in general appeared fruitless.

These attempts failed as a result of the proposition of the Arab States and their supporters, that an act of terrorist violence committed in the name of some national ideological cause is a form of justifiable homicide that deserves the indulgence of the international community. It is regrettable that propositions like the above can be used and are being used to torpedo a proper application of the Declaration on non-invention 1965 (see also Appendix on page 88).

With one or two exceptions, South Africa was spared serious terrorist interference and the situation was changed drastically by the application of the policy of separate development. This development will render the defence against terrorist interference more and more a matter for the authorities of the national homelands. The UN will have to take this into account when it is confronted with proposals made in its meetings to side with persons who—having committed subversive, terrorist and armed activities under the pretence of wishing to liberate South Africa or to combat apartheid—are subsequently caught. If ever SWAPO, the South West African Peoples Organization, (to be referred to lateron), should develop such activities in South West Africa, South Africa completing its transfer of power may be expected to assist the peoples in the area to combat those activities. It could not be expected that the Security Council in such case would decide to condemn any self-defense of these peoples.

In so far as the above predictions are based on the Declaration on non-intervention 1965, one should not ignore the incompatible resolution passed by the Security Council on 23 July 1970 (Resolution 282 (1970)). The Declaration states that all States shall respect the right of self-determination and independence of peoples and nations, to be freely exercised without any foreign pressure, and with absolute respect for human rights and fundamental freedoms. One would say that consequently not only the South African Government but also the various peoples of South Africa which are in the process of realising their rights of self-determination and independence, could appeal to this Declaration in order to safeguard themselves against outside intervention for any reason and to defend themselves if necessary by force of arms. But such thoughts are completely incompatible with the Security Council Resolution 282 (1970) which states that the situation resulting from the continued application of the policies of apartheid and the constant build-up of the South African military and police forces constitutes a potential threat to international peace and security. The resolution furthermore recognises that the extensive arms build-up of the military forces of South Africa poses a real threat to the security and sovereignty of independent African States opposed to the racial policies of the Government of South Africa, in particular the neighbouring States.

France, the United Kingdom and the USA abstained from voting on this resolution and one can imagine their concern for the effect which it could have on the maintenance of international peace and security in Southern Africa. Whilst the salutary effect which this resolution could have on international peace and security remains obscure, the opposite effect is easy to predict. The neighbouring States of South Africa could interpret the resolution as a promise of support in case they should become engaged in an armed conflict with South Africa when opposing the racial policies of the South African Government which would force the latter to take up arms to safeguard its own internal security. Supposing that such opposition should take the form of a neighbouring

State's sending armed terrorists into South African territory, then a resolution of this nature could be interpreted as meaning that South Africa would not be allowed to defend itself inasmuch as such action might constitute a threat to the security and sovereignty of the neighbouring State in question. What purpose is served by this Security Council Resolution? That South Africa has an equal right to preserve its own sovereignty intact is completely ignored!

What if a terrorist invasion from a neighbouring State should assume the dimensions of an armed attack, which South Africa would under the Charter be fully entitled to repel? 'Nothing in the present Charter shall impair the inherent right of individual or collective self-defense if an armed attack occurs against a Member of the United Nations, until the Security Council has taken measures necessary to maintain international peace and security' (art. 51 Charter). We can only hope that South Africa's neighbouring States will be guided in future more by the clarity of the Charter then by resolution 282 (1970) of the Security Council, which is a bad compass for those earnestly wishing to set a course towards international peace and security.

X

# The progress of the transformation of the legal order in South Africa

The stand taken by the UN in regard to terrorism, forces one to the shocking conclusion that there is something horribly wrong with the world's conscience, in so far as such a conscience still exists and is still able to make itself felt, for certainly the world's reaction to this phenomenen which we call terrorism is no longer what it was prior to World War II. The fact that terrorism is openly condoned and even applauded in many countries in Africa and elsewhere, is having far-reaching consequences, especially for South Africa.

UN intervention in South Africa's affairs has developed into a question in regard to which the Security Council is 'exercising the functions assigned to it in the United Nations Charter'. South Africa is to be given time to find an alternative for its obsolete political order and to develop it: 'a positive alternative to be found through peaceful means', as required of South Africa in terms of S.C. Resolution 182 (1963) of 4 December 1963. Since then, many years have passed, and during this period South Africa has gone a long way towards meeting this requirement. The task has not been an easy one, given that country's peculiar circumstances and the nature of the problems South Africa faces. In particular in the task of accelerating economic development, and so moving closer towards self-determination for each of South Africa's peoples, it is to be hoped that the Western powers will in the near future start offering economic and technical aid to help speed up this process.

What is clearly already happening is that, in a natural and logical way, effect is being given to the Security Council's recommendation following the report of the Group of Experts, viz. that 'all the people of South Africa should be brought into consultation and should thus be enabled to decide the future of their country at the national level' (Resolution 191 (1964) of 18 June, 1964).

This consultation is proceeding on the basis of granting self-determination to the peoples qualifying for it in terms of the purposes of the United Nations Organization.

The Governments of Transkei and the other emerging states opting for self-determination and independence are engaged in a process of consultation with each other as well as with the sovereign Government in Pretoria. The future interrelationships are being discussed with earnest intentions further to adapt to each other and to practise tolerance and live together in peace with one another as good neighbours, in accordance with the principles of the UN Charter. Racial aspects, as e.g. those of a 'black' or 'white' (majority) government or discrimination on the social levels below the governmental level within the legal territory of one of those parties, will not be acceptable to the other-ones. The current consultations are taking place amongst parties which understand each other very well on these points and the UN would do well to take due note thereof.

The UN should in this respect seek to establish and maintain honest contacts with the Governments of Transkei and the other emerging states so as to be able positively to participate in this development.

An observer could predict that the interference of the UN in the affairs of South Africa should cease at a point in time when the policy of separate development is fully realised—which would be at the time when the modernised legal order would have banned all despicable discrimination—or so much earlier as the General Assembly would accept that the Security Council is not prepared further to confront South Africa.

From a UN point of view it would appear difficult forthwith to put a stop to the campaign against apartheid, after having decried the policies of apartheid practised by the Government of the Republic of South Africa as abhorrent to the conscience of mankind for so many years. It would furthermore be more than difficult to scrap the costly apparatus created to conduct and sustain that campaign and to aid the victims or pretended victims of apartheid, even when the redundancy of this apparatus is realized.

The serious agitation by the Churches and other anti apartheid organisations which had so much credit in the UN-sphere will not easily cease and will continue to demand reactions by this Body. No observer can easily predict how the UN will find its way out of these problems, or how it will overcome the tensions within its organisation possibly caused by these problems.

Perhaps South Africa itself could render assistance to ease these tensions by way of agreeing to the creation of a Court of Justice whose main task would be to guarantee and safeguard human rights in Southern Africa. This idea was already advanced by the Group of Experts formed by the Secretary-General of the UN to implement Resolution 182 (1963) of 4 December 1963 of the Security Council. It therefore emanated from the United Nations itself. This proposed Court could be created with the cooperation of all Southern African States concerned—including Transkei and the other emerging national homelands—and it could cover the whole of Southern Africa. It could contribute greatly to the policies of détente of the leading Statesmen, which now is going on in that part of the world.

For the implementation of such a scheme, South Africa and Zambia and the other nations of Southern Africa might wish to obtain the accord of the General Assembly. Were the General Assembly to agree to this, such a step would herald a return to 'a conciliatory approach in accordance with the principles of the Charter', which, prior to 1960, still had the approval of the majority in the General Assembly and was advocated in G.A. Resolution 1016 (XI). This would require a radical change in the attitude of the General Assembly, which during the period 1969-1975 has been growing more and more intransigent making ever more onreasonable demands on South Africa, with scant regard to the fact that these demands have deviated ever further from the decisions of the Security Council relating to South Africa.

It will not be easy to effect a change in the attitude of the General Assembly as long as it sets its compass on the Special Committee's course. The composition of this committee is extremely onesided as none of the Western democracies is represented in it.

It fails completely to provide the General Assembly with objective reports and information which could be the basis of an honest and fair appraisal by this body. In accordance with its task to conduct a 'campaign', it publicizes reports aimed at putting the South African Government in the wrong and which at times are downright misleading. As an example may serve the objections raised in the 1975 report A/10022 against the socalled 'Bantustan' policy (para 78) and against the legal transformation which the South African Government has been and still is effecting.

This report A/10022 as well as earlier reports are quite the opposite of a conciliatory approach and since as such they are of no use to the Security Council, which furthermore lends no authority to them, they might hinder efforts to coordinate the policies of the General Assembly and the Security Council.

XI

# Security Council and General Assembly within the overall system of the UN Organization

One fact which South Africa—and also anyone venturing upon prognoses in regard to future UN intervention in South Africa's affairs—has to take into account, is that the great majority of the Members of the Security Council are no less implacable in their attitude towards South Africa than the majority in the General Assembly.

The minority in the General Assembly, i.e. those powers who are opposed to UN measures such as economic boycotts for the purpose of enforcing a transformation whereby the South Africans of European extraction would find themselves in the hopeless position of a powerless minority, has become reduced in numbers to include no more than the Western European countries and the USA. These happen to be the countries most concerned about the fate of white South Africans, in the same way that India is concerned about the lot of people of Indian origin domiciled in South Africa and in other African States. These powers also take a different view of terrorism from that adopted by certain African states.

The situation in the Security Council is, briefly, as follows: The three permanent members, France, the UK and the USA, are continually under pressure from the other members, or at least the majority of other members, to take up a position which is unfavourable to South Africa, i.e. a decision to force it to withdraw prematurely from South-West Africa ('Namibia'), and to weaken its powers of self-defence to the point of rendering it incapable of repelling terrorist incursions.

The Security Council is, however, bound by the provision of the Charter that decisions on matters of the kind under discussion require the 'affirmative vote of nine members, including the concurring votes of the permanent members'. The resolutions passed by the Security Council with regard to the internal policies of the Government of South Africa in 1960, 1963, 1964 and 1970, are indicative of the limits to which the Security Council will be prepared to go in applying coercive measures.

As far as France and the United Kingdom were concerned, these limits were exceeded by the Council when it passed Resolution 134 (1960) of 1 April 1960, condemning the alleged 'situation arising out of the large scale killings of unarmed and peaceful demonstrators against racial discrimination and segregation in the Union of South Africa', and Resolution 181 (1963) of 7 August 1963, noting with concern 'the arms build-up by the Government of South Africa, some of which arms are being used in furtherance of that Government's racial policies', and requiring the South African Government to liberate all persons imprisoned, interned, or subjected to other restrictions for having opposed the policy of apartheid'. France and the UK abstained from voting on these resolutions.

In regard to Resolution 190 (1964) of 9 June 1964, which constituted interference in and condemnation of the Rivonia trial, the USA joined with the other two permanent members, France and the UK, in withholding her support (so. incidentally, did Brazil). The same applied to Resolution 282 (1970) of 23 July 1970, which proclaimed a far-reaching arms embargo, and declared 'that the extensive arms build-up of the military forces of South Africa' posed 'a real threat to the security and sovereignty of independent African states opposed to the racial policies of the Government of South Africa, in particular the neighbouring states'.

A study and analysis of the import of these resolutions gives a fairly clear idea of the extent to which the three Western permanent members will allow intervention by the Security Council. They may tolerate the adoption of resolutions but decline to assume co-responsibility; and from this it may be concluded that they will not countenance intervention in the form of an economic boycott or intervention in any form which might be construed as being contrary to the proviso concerning 'peaceful means' in the Council's Resolution 182 (1963) of 4 December 1963, even though the General Assembly had for many years past been clamouring vociferously for such intervention.

In view of the fact that the majority of the members of the Security Council have undoubtedly done everything in their power to persuade the Western permanent members to yield to their urging, but without any notable success, it is fairly safe to assume that these members would veto any draft resolutions to this effect.

It is further apparent that these permanent members have never yet supported any decision condemning the transformation of South Africa's legal order on the basis of self-determination for each of its various peoples, a system which has consistently been rejected by the majority in the General Assembly. Finally, the view expressed by the General Assembly that 'freedom fighters' in South Africa who have been imprisoned for committing acts of terrorism 'should be treated as prisoners of war under international law, particularly the Geneva Convention relative to the 'Treatment of prisoners of War of 12 August 1929'' ' (G.A. Resolution 2396 (XXIII) of 2 December 1968), has never been incorporated in the decisions of the Security Council.

In regard to the tolerance shown by the Western permanent members, it should be remembered that this tolerance—already referred to earlier—was probably strained to the utmost by the other members in July 1970, when they insisted on passing Resolution 282 (1970), in which it was made clear that should any situation arise in South Africa as a result of the activities of 'freedom fighters', for which a neighbour state could be held accountable, the Council would unjustly side with the latter, unjustly, in view of the fact that in terms of the UN Charter, this organization is responsible for ensuring and safeguarding the peaceful co-existence of peoples.

In the years after 1970 the General Assembly continued its crusade against South Africa in complete disregard of the consideration that the Security Council would withhold its indispensable support as well as that the crusade could possibly encourage terrorism within South Africa.

A divergence in points of view and policy towards South Africa emerged between the General Assembly and the Security Council but it remained unclear to what extent the Security Council would be succesful in keeping the General Assembly from decisions which would isolate, harm and humilitate South Africa: decisions which in any case would remain mere slogans as the Security Council would not endorse their implementation.

In the years 1972, '73 and '74 many decisions were taken—upon the recommendations of the Special Committee on apartheid—which had the effect of isolating, harming and humiliating South Africa in disregard of the limit of the UN's authority, which is the sovereignty of States. This limit was underscored during the General Assembly debate on the subject 'Strengthening of the role of the United Nations with regard to the maintenance and consolidation of international peace and security etc' on 6 December 1974 (A/PV. 2307) by the delegate of France, Mr. de Guiringaud: 'I said earlier, that we are here as if in a parliament without being a parliament. The General Assembly, in particular can adopt only recommendations. It cannot legislate '. . . the activities of the Assembly are part of the overall system within which another body, the Security Council, exercises essential responsibilities and prerogatives. Our Charter reserves to the Council the primary role in regard to the admission of States and the expulsion or suspension of Members of the United Nations. The Assembly must take care not to ignore these rules in exercising, on the basis of a competence which in fact is limited, the rights and privileges which the Charter gives to each of the Members of our Organization. The Assembly would be doing itself an injustice if it were to betray the letter and the spirit of our basic laws in these regards'.

The General Assembly's decisions in the years 1972-74 concerning South Africa completely ignored the limitations imposed by the sovereignty of States, but they also failed to appreciate the essential responsibilities and prerogatives of the Security Council. As examples of these decisions, unacceptable for each individual UN Member as well as for the Security Council, we may cite:

1. the 'condemnation' of the establishment by the 'racist' Government of South Africa of the so-called Bantustans and of the relevant reallocations of domiciles, rejected as 'prejudicial to the territorial integrity of the country and the unity of the people' (Gen.Ass.Res. 2933 E (XXVII) of 15 Nov. 1972; 3151 G (XXVIII) of 14 Dec. 1973 and 3324 E (XXIX) of 16 Dec. 1974). In the latter resolution mention is made of 'the manoeuvres of the South African régime to perpetuate apartheid by establishing "Bantustans" ';

2. the 'reaffirming' of 'the legitimacy of the struggle of the oppressed people of South Africa to eradicate apartheid and racial discrimination by all available means and to attain in the country as a whole majority rule based on universal suffrage' (Gen.Ass.Res. 2933 E (XXVII));

3. the 'declaration' that 'the South African régime has no right to represent the people of South Africa and that the liberation movements recognized by the Organization of African Unity are the authentic representatives of the overwhelming majority of the South African people'. (Gen.Ass.Res. 3151 G (XXVIII)).

These resolutions deny the authority of the sovereign South African Government as well as the authority of the Governments of the national homelands to lead their peoples to self-determination.

It is further worthy of note that the resolution of 16 December 1974 challenged directly the authority of the Security Council which on 30 October 1974 decided to reject the proposal of Kenya, Mauritania and the United Republic of Cameroon (see Appendix 3 page 91) to recommend the expulsion of South Africa from the United Nations. The relevant passage reads: 'noting with concern that three Powers, permanent members of the Security Council, namely France, the United Kingdom of Great Britain and Northern Ireland and the United States of America, by the use of the veto, prevented the Council from taking effective action against the South African apartheid régime'. This resolution of 16 December 1974 was however virtually no more than a post mortem: as the General Assembly had in fact already annulled the decision of the Security Council, a decision which will be discussed in the chapter XIII. In the General Assembly plenary meeting of 12 November 1974 President Bouteflika (Algeria) applied the rules of procedure in such a way that the General Assembly refused to allow the delegation of South Africa to participate in its work.

This course of action was disputed, particularly by the spokesmen of the three abovementioned permanent members of the Security Council, on the basis of arguments derived from the Charter, thus refuting the reasoning of the President which was based on an interpretation of the rules of procedure (art. 29) but had led to a result inconsistent with the provisions of the Charter itself.

The General Assembly endorsed the ruling of the President with a vote of 91 in favour, 22 against and 19 abstentions. Thus the General Assembly reached a decision in breach of a prerogative of the Security Council. Decisions affecting the membership of the United Nations can only be taken upon the recommendation of the Security Council (Articles 4-6 Charter). In this case the Council did not fail to make or postpone a recommendation, but it had emphatically refused to do so on 30 October 1974. It was then that South Africa's position was consolidated and protected in the face of pressures and machinations of the General Assembly which for years had been trying to deprive this nation of its rights and privileges as a UN Member. One of these rights was the right to participate in the work of the General Assembly.

The Security Council is not expected to acquiesce in the undermining of the legal order based on the UN Charter and consequently of South Africa's legal position and it will have to search for ways and means to redress the damage: not only in the case above referred to, but also in all those previous instances where General Assembly resolutions of the years 1972-1974, mentioned earlier, damaged the legal status of South Africa.

# XII

# The arms embargo's

As was explained in the foregoing, all infringements upon the rights of South Africa as a sovereign state and UN member by the General Assembly, in whatever manner dressed up and formulated, emanate from the overall policy of the General Assembly to prevent that South Africa modernises its legal order in accordance with the policy of separate development.

The 'declaration' of 14 December 1973 that 'the South African régime has no right to represent the people of South Africa and that the liberation movements recognized by the Organization of African Unity are the authentic respresentatives of the overwhelming majority of the South African people' (Gen.Ass.Res. 3151 (XXVIII)) was no less than an excessive and spectacular implementation of the overall policy. If that policy should be abandoned, the General Assembly would no longer have a basis to deviate—in support of the liberation movements—from the principles to which it committed itself in its Declaration on the Inadmissibility of Intervention in the Domestic Affairs of States and the Protection of Their Independence and Sovereignty of 21 December 1965, General Assembly resolution 2131 (XX), of which the full text appears as an appendix on page 88.

By its declaration of 14 December 1973 the General Assembly damaged prospects for the maintenance of international peace and security in Southern Africa. For it meant an encouragement to the liberation movements to do battle with the South African Government, with the authorities in the national homelands, with each other and with any new liberation movements yet to emerge. Should South Africa be too weak to counter such developments, chaotic situations such as those witnessed in Angola in 1976 could arise.

To this eventuality is clearly related the question whether South Africa should preserve the possibility of buying arms abroad. The General Assembly desired to cut off the arms supply and has repeatedly recommended an arms embargo. Its purpose would be to provide an instrument in the implementation of its overall policy as described above. The Security Council which did not endorse such overall policy did in fact allow itself to be persuaded by the General Assembly in 1970, to recommend an arms embargo on its own part (Security Council Resolution 282 (1970) of 23 July 1970). France, the United Kingdom and the United States however withheld their affirmative vote. Later on the Security Council was again confronted with the overall policy of the General Assembly against South Africa, which it rejected in respect of the instrument of expulsion of South Africa from the UN (to be referred to later), as well as of the instrument of a mandatory arms embargo. The latter was incorporated in a proposal which was rejected by the Security Council on 6 June 1975 during the closing stages of a debate on Namibia (South-West Africa) to which we intend to revert later.

If the Security Council really desires to call a halt to a continuation of the General Assembly's aforesaid overall policy, (because such a policy would lead to chaotic situations in South Africa which could endanger international peace and security), it would seem more logical for the Council to recommend that the South African government be assisted to accelerate the implementation of its policy of separate development and that the governments of the national homelands be armed to repel attacks by liberation movements, rather than to maintain its arms embargo which it recommended on 23 July 1970.

It would seem obvious that the prominence given by the OAU and the General Assembly to liberation movements which would depose the South African authorities, is not appreciated by the three permanent members France, UK and USA as they likely view it as a stratagem aimed at further endangering peace and security in Southern Africa.

These three permanent members, who withheld their affirmative vote on the abovementioned Resolution 282 (1970) in the Security Council on 23 July 1970, might have been expected to have made use of their right of veto, if they could have foreseen at that time the Cuban-Russian invasion of Southern Africa (Angola) during early 1976.

XIII

# The hour of truth (30 October 1974)

From 18 to 30 October 1974 the Security Council debated the proposal of the General Assembly laid down in its resolution 3207 (XXIX) of 30 September 1974 to review the relationship between the UN and South Africa in the light of the constant violation by South Africa of the principles of the Charter and the Universal Declaration of Human Rights. On that occasion pressures were brought to bear on the Security Council to render its indispensable assistance to bring about the expulsion of South Africa from the UN.

The debate was remarkable in that it contained hardly any discussion of South Africa's own solution to end the obsolete racial relationships and the discrimination by way of 'separate development'. This matter was only briefly touched upon during the beginning of the debate by the spokesman for Somalia, Mr. Galib (S/PV 1796, page 22). 'The independent African States'—according to this spokesman—'by reason of experience and kinship and by our geographical and cultural ties, and our responsibilities towards our African peoples in South Africa, fully understood the true aspirations and the birthright of the majority of the people of South Africa.' In the first part of this statement it was recognized that South Africa contains within its borders several peoples. Where reference is made to the aspirations of the majority of 'the people' of South Africa, the statement fails to recognize the fact that the various peoples may each have their own aspirations. Without further identification it remains unclear to which people of South Africa reference is made, nor is it clear what aspirations are being referred to. 'We know'—so Mr. Ghalib added—'that the Bantustan scheme is a monumental fraud perpetrated at the expense of the African people. It was conceived only to isolate them from each other, to divide in order to rule and to keep Africans from the mainstream of the national life of the territory.'

In the order of the above mentioned debate, the above statement dit not need to be refuted. Other countries are free to prefer another scheme than that which is being applied in South Africa, but that is no more than a political preference, which, as such, is not accepted in South Africa in the exercise of self-determination by the peoples of South Africa.

Since this essential point was not further discussed in the extensive UN debate, 'apartheid' could be critized and condemned, but only in an essentially abstract manner, i.e. completely divorced from the reality of the solution of separate development. Under such conditions, all countries including the Western Powers could associate themselves with such condemnation, but the preparation of decision-making remained a somewhat obscure affair in sofar as the Western Powers were concerned, as these powers would not attach a condemnation in concreto to the condemnation in abstracto.

The spokesman for Great Britain especially criticized South Africa in a way which was incompatible with the reality of the development of South Africa. 'We are quite convinced'—said Mr. Richard of the United Kingdom—'that the Government of South Africa must change its policies of racial discrimination. We have worked for that change and we will continue to work for it. We will use whatever influence we have to that end.' (S/PV. 1808 page 41). This statement completely ignored the enormous efforts of the South African Government to find a solution through which there would no longer be any place for all those racial discriminations which were part and parcel of the governmental structure which, it should be noted, was inherited from the British Government.

On the other hand the British spokesman did draw the attention to the abstract character of the debate. 'I must say that this debate has become more of an exercise in condemning South Africa than a review of the relationship between South Africa and the United Nations.' The crucial point for his rejection of the proposed expulsion of South Africa from the United Nations, was found by the British spokesman in the plea for the alternative process of conciliation and the application of peaceful means for the solution of problems as set out and recommended to Members of the Organization in its Charter.

The proposal to render assistance in the expulsion of South Africa from the UN (S/11543 see appendix 3) was rejected by the Security Council on 30 October, 1974. France, the United Kingdom and the United States of America cast a negative vote by making use of their right of veto. The hour of truth for the position of South Africa in the UN in terms of her dependence on the UN Charter and of the authority granted by that Charter to the Security Council and within that Council to the Permanent Members, had come.

This solemn occasion was somewhat spoiled by the observation by the President, Mr. Njine of Cameroon, who declared after the voting: 'In the opinion of my delegation the status of South Africa in our Organization is quite clear, inasmuch as it is a Member now only because of the veto power'. By this statement the authority held by the Council was rendered problematical. Against this very point a warning voice had been raised by the British representative, Mr. Richard, when he declared: 'There is, however, a quasi-constitutional argument which has crept in, that it is somehow the function of the Security Council to recommend expulsion because that course is assumed to be the will of the vast majority of the General Assembly. But the Council cannot be absolved from reaching its own decision. Indeed, that is precisely what the General Assembly has asked it to do. Moreover, that is what the Charter specifically provides for.'

On 30 October 1974 the Security Council expressed itself in no uncertain terms—and in accordance with article 27 of the Charter which governs the voting procedure in the Security Council—against the proposal to recommend the expulsion of South Africa from the UN Organization. Nor did it wish to yield to the pressures which the General Assembly for so many years had been bringing to bear upon the Security Council in order to have it recommend expulsion in terms of the latter's prerogative as laid down in article 6 of the Charter.

The President of the Security Council in his letter to the President of the General Assembly of 31 October 1974 (A/9847) merely reported 'that the Security Council was not able to adopt a resolution on this item and accordingly remains seized of the matter'.

It would have been clearer if this report had not left the ability of the Security Council problematical. For in fact its ability was demonstrated on 30 October 1974. The President of the Security Council could well have observed that the General Assembly had not been able to convince the Security Council, rather than express his personal regrets on the matter.

In terms of a proper functioning of the Charter it would furthermore have been appropriate for the General Assembly to accept reality, rather than to consider a continuation of its pressure upon the Security Council which would be tantamount to causing a tug—of—war between the two organs in a manner incompatible with the division of tasks and responsibilities as laid down in the Charter.

It is simply inconceivable that the General Assembly will continue in future to try to persuade the Security Council to use its authority to demand that the Government of South Africa abandon its policies of separate development and institute a unitarian state in which the population could be distinguished and divided into a black majority and a white minority.

In the Special Political Committee in October 1975 no objections were raised against the statement that the responsibility for the future of the South African population as well as for parts of the population rests solely with the South African Government and that this responsibility is not shared by the United Nations as was generally accepted in the case of colonial territories in the past. South Africa is a sovereign State and a Member of the United Nations. 'The formal assumption of a special responsibility by the United Nations for parts of the population of the Member State was not envisaged in the Charter and would exceed the Organization's authority.' Thus Mr. von Uthmann (Federal Republic of Germany) in the Special Political Committee on 28 October 1975 (A/SPC/SR.964).

Similar views about the extent of UN authority were expressed during that same meeting by spokesmen of the United Kingdom and France.

The political tug-of-war between the General Assembly and the Security Council can never be expected to result in a situation whereby the General Assembly will succeed in drawing the Security Council beyond the limits of the Organization's authority!

# XIV

# The United Nations and South West Africa (Namibia)

International attention given to South Africa both at the UN and more generally has increasingly been directed at the South-West African or (since 1968) Namibian issue. This question, as a facet of South Africa's foreign relations, is a matter which justifies fuller treatment.

South Africa, as one of only three African members of the League of Nations, was given the mandate over South West Africa—a territory which it had militarily occupied on behalf of the United Kingdom at the outbreak of World War I. The mandate determined that, as a Class C mandate, South West Africa could be administered 'as an integral portion of South Africa'. The fact that South Africa was granted such international responsibilities is an indication of the international esteem which it enjoyed at that time.

When the league of Nations was dissolved in April 1946, the question arose of the future of South-West Africa. Among the various courses which suggested themselves, were the following: (1) The territory be placed under international trusteeship; (2) It be annexed by South Africa; (3) It become independent; and (4) The prevailing situation be maintained. Even before the dissolution of the League, the Legislative Assembly of South West Africa had asked that the territory be incorporated in South Africa. And this was the request which General J. C. Smuts, South African Prime Minister of that time, made to the UN in 1946. But this request was refused and South Africa was asked to transfer the territory to the new Trusteeship Council of the UN. South Africa declined to do so, adopting the position that she was under no legal obligation to the UN to do so. By 1947 it was clear to South Africa that the UN would never agree to incorporation and in June of that year the government informed the UN that it had no intention of pursuing this course: it would continue to administer the territory in the spirit of the mandate and it undertook to furnish the UN with voluntary progress reports.

However, the new government in 1948 decided to discontinue this practice. of reporting. In consequence the UN asked the International Court of Justice for an opinion of the legal status of South West Africa and for a statement of South Africa's obligations under the mandate. This opinion was handed down on 11 July, 1950; further clarification was sought from the Court in 1955 and 1956.

Following the 1950 opinion, the UN suggested the establishment of a so-called Committee on South West Africa which was intended to liaise between the inhabitants of South West Africa and the UN until an agreement was reached. But with no settlement in view, the General Assembly in 1957 recommended further recourse to the International Court, and subsequently Ethiopia and Liberia (as original members with South Africa of the League) instituted proceedings. After a lengthy trial the Court held, in a judgement handed down in 1966, that the applicants hand failed to establish 'any legal right or interest appertaining to them in the subject-matter of their claims'.

The General Assembly responded swiftly and drastically to this judgement when, on 27 October 1966, it revoked South Africa's mandate over South West Africa and called upon South Africa to withdraw from the administration of the territory (Resolution 2145)—a resolution which was described as illegal by the representatives of France, the United Kingdom, South Africa and other countries. Resolution 2145 also called for the establishment of a Council for South West Africa, consisting of fourteen member-states and charged with the task of 'recommending practical means by which South West Africa should be administered, so as to enable the people of the territory to exercise their right of self-determination and to achieve independence'.

The Security Council confirmed the General Assembly's revocation of the mandate by Resolution 264 of 20 March 1969 (France and the United Kingdom abstained), and this was re-affirmed by Resolution 269 of 12 August 1969 (Finland, France, the United Kingdom and the United States abstained). It should be pointed out that South Africa responded to both these resolutions in some detail. Essentially, however, it questioned 'the legal basis for the activities of the so-called Council for South West Africa or for Security Council intervention'.

Pursuant to Resolution 276 (1970) of 30 January 1970 the Security Council established an *ad hoc* sub-committee to consider ways and means by which its decisions could be effectively implemented. And this sub-committee subsequently recommended *inter alia* that an advisory opinion be obtained from the International Court of Justice. When the Security Council again considered the question it passed two resolutions. In the first, Resolution 283 of 29 July 1970, it requested (with France and the United Kingdom abstaining) all states to desist from any action or any form of relations with South Africa which might imply recognition of the authority of the South African government over the territory; and in the second, Resolution 284, the Security Council referred the dispute back to the International Court (Poland, the Soviet Union and the United Kingdom abstained).

The Court handed down its opinion on 21 June 1971. It held that the mandate had been lawfully terminated (which was the more important question put to it); and it then went on to spell out at some length the legal consequences for states of the continued presence of South Africa in South West Africa, notwithstanding the Security Council's Resolution 276 (1970), which was the other question posed to the Court. In this regard its opinion was that, as South Africa's presence in South West Africa (or Namibia) was illegal, it was obliged to withdraw immediately. And secondly, that all member-states of the UN were under an obligation to heed the illegality of South Africa's position in South West Africa.

An interesting feature of the 1971 hearings was a formal request by South Africa that a plebiscite be held in South West Africa in order to determine whether the local population wished to be governed by South Africa or the UN. Although South Africa suggested that the Court supervise the plebiscite jointly with the South African government, the Court nevertheless rejected this proposal.

With Resolution 301 of 20 October 1971, the Security Council adopted the Court's advisory opinion (France and the United Kingdom abstained). One of the provisions of Resolution 301 was the request to the Secretary-General that he report periodically on the implementation of this resolution, and this was expanded by Resolution 309 which was adopted at a special meeting on African problems of the Security Council held in Addis Abeba in February 1972. The Secretary-General was authorised to enter into discussions 'with all parties concerned on the future of South West Africa'. During March 1972 Dr. Waldheim, at the invitation of the South African government, paid a five-day visit to that country and to South West Africa. After receiving the Secretary-General's report the Security Council by Resolution 319, adopted on 1 August 1972, authorised him to appoint a personal representative to assist him. Following consultations with South Africa, Dr. Alfred M. Escher, a retired ambassador for Switzerland, was named.

Dr. Escher subsequently visited South Africa and South West Africa from 8 October to 3 November. And in his report to the Secretary-General he urged that further contact with the South African government be maintained. This was endorsed by the Security Council in Resolution 323, adopted on 6 December 1972. The resolution required the Secretary-General to report on progress made by the end of April 1973. However, the Secretary-General's report was only considered by the Security Council in December of that year, and by Resolution 342 of 11 December 1973, the Security Council decided to discontinue further contacts with South Africa as provided for by Resolution 309.

In a statement issued on 12 December 1973, the South African Minister of Foreign Affairs expressed the view that the talks which had taken place had led to positive results, and that even Dr. Waldheim had acknowledged this. However, the so-called Council for South West Africa (or Namibia) as well as the UN Committee of Twenty-Four (that is the Decolonisation Committee) 'had gone out of their way to condemn these efforts in advance and to oppose anything positive which might flow from the discussions'. The South African Foreign Affairs Minister concluded his statement as follows:

> Whatever happens, South Africa firmly adheres to its attitude—which in the past has been repeatedly expressed—that it is for the inhabitants of South-West Africa to decide their own future. Neither the South African government nor the UN should enforce solutions upon them from the outside. In the meantime the South African government will continue with its efforts to assist them on the way to self-determination. In spite of the latest developments in the Security Council, it is basic to the Government's policy to exchange ideas with others who have the real interets of the inhabitants at heart.

On 17 December 1974, the Security Council adopted an African-sponsored resolution (numbered 366) which gave South Africa until 30 May 1975 to withdraw from South West Africa. South Africa having failed to comply, the Security Council was called into session on 30 May 1975 and deliberated until 6 June 1975. A proposal was made to the Council to take measures under Chapter VII of the UN Charter and to declare that the illegal occupation of the territory of Namibia by South Africa constituted a threat to international peace and security. It was also proposed to impose a mandatory arms embargo in its broadest sense against South Africa (S 11713 - see appendix 4). France, the

United Kingdom and the United States of America cast their vote against this draft resolution on 6 June, 1975, whilst Japan and Italy abstained from voting. The President noted that: 'The draft resolution accordingly was not adopted owing to the negative votes of three Permanent Members'. This was the second time in the history of the UN that a draft resolution was rejected by a triple veto in the Security Council.

It would not be an exaggeration to observe that thus in the UN the hour of truth for the fate of South Africa had come for the second time. It is, in fact, not the case that the Security Council adopted the view that South Africa had not fully followed the recommendations of the earlier Resolution 366 of 17 December 1974. But it is the case that according to the statements of the spokesmen for the Western Permanent Members, the notion was rejected that South Africa, by non-observance of the directives of Resolution 366 in her treatment of the problems of Namibia, had created a situation or allowed a situation to continue, which could be qualified by the Council as a threat to international peace and security (and which therefore could be the basis for action under Chapter VII of the UN Charter).

This interpretation of the Charter by the Security Council blocks, in terms of the Charter, all means to exert pressure which the General Assembly would like to apply against South Africa in order to force that country to abandon its scheme for the realization of self-determination of the several peoples, which live side by side in Namibia, in favour of a one man-one vote system based on the fiction that the Namibian territory is inhabited by only one people.

Preference for an other system, based on that fiction, was voiced by the spokesmen of the Western Powers, but only in terms of recommendations to find a solution through negotiation, reconciliation and other means as described in article 33 of the Charter.

The French spokesman recommended that South Africa should not refuse to consider the adoption of a perfectly democratic and simple system which consists in appealing for the vote of the entire population concerned. The future political structure, so he suggested, would after the election of the constituent assembly be defined by the elected representatives of the population (S/PV 1824 page 56). This recommendation however failed to recognize that in that case smaller peoples (e.g. the Herero) would numerically be represented in the constituent assembly so weakly, that they would not be able to oppose a political structure in which there would be no place for them to exercise their right of self-determination.

The British spokesman said a.o.: 'I should perhaps add that my own government makes no secret of its preference for a single Namibian State, though it is of course content to leave this matter—as it should be left—to the wisdom of the people of Namibia themselves'. He appeared satisfied in his own mind that the South African government would fulfil its promises in that respect.

# XV

# Self-determination for South West Africa

The promises by the South African Government were founded upon the reality of South West Africa being inhabited by a number of peoples living far apart, whose only contact consisted of waging war, who never came to understand one another, and who, each for themselves, opted for self-determination.

Out of a total population of 760 000 souls the individual peoples form minorities against the Ovambo- and Kavango peoples living in the North, which together account for 53% of the total. The Herero (6.6%), the Nama (4.3%) and others will never subject themselves to the more numerous peoples. The same applies to the Rehoboth Bastards, the Tswana, as well as the Damara. All these peoples, excepting the Damara (8.7%) sent delegates to a conference to discuss the constitutional future of South West Africa, held at Windhoek during 1-12 September 1975. What was involved was a constitutional conference, which was enabled to succeed due to the participation of a representative Ovambo-delegation, which could be formed as a result of the Ovambo-land elections of January 1975. Fiftyfive percent of the population took part in these elections in spite of fierce efforts by SWAPO to persuade the voters to stay away from the polls.

After the opening of the conference each delegation appointed two representatives to a credentials committee, which reportedly approved all 11 delegations as 'properly representative'. The conference reached unanimity on a 'declaration of intent' concerning independence, which envisages a three year period in which to achieve independence of S.W.A. or Namibia (full contents of their declaration appear in Appendix no. 5). The prospective political structure will be a confederation on ethnic base. Interference by the UN 'and all other outsiders' was rejected and solutions by forcible means were condemned.

The conference hereby took a stand clearly opposed to the above-mentioned SWAPO (South West African Peoples Organization), which the UN Council for Namibia considers the only representative organ for the peoples of Namibia.

The UN Council for Namibia wasted no time in submitting a report to the General Assembly recommending that strong pressure be brought to bear upon South Africa because of its continuing objections to withdrawing from Namibia and because of its efforts to consolidate its illegal occupation of the country by organising a so called constitutional conference aimed at creating a division amongst ethnic groups and to perpetuate its Bantustan policies.

It is clear that this recommendation would gladly have been followed by the majority of the members of the General Assembly but not by the Western nations which would not easily accept a fictitious self-determination, i.e. a self-determination of a fictitious 'people', the identity of which could only be determined by the borders of neighbouring countries. The controversy in viewpoints was manifested in a resolution adopted by the General Assembly on 6 November 1974, G.A. 3222 (XXIX). This resolution reaffirmed the rights of all peoples, in particular the peoples of Southern Africa, to self-determination, freedom and independence and strongly condemned the policy of those States-members of the North Atlantic Treaty Organization—and of other powers which are assisting the racist régimes in Southern Africa and elsewhere to repress the profound aspirations of peoples to the enjoyment of human rights and to prevent the exercise of those rights.

If a majority in the General Assembly should accept a resolution following the above-mentioned recommendations of the Council for Namibia, it may be expected that South Africa, upon the recommendation of the peoples who participated in the constitutional conference, will ignore such resolution and that furthermore the Security Council will not be prepared to enforce the G.A. resolution against South Africa. Such events will be the result of a difference in judgement of the profound aspirations of peoples to enjoy self-determination.

The above mentioned G.A. Resolution 3222 (XXIX) of 6 November 1974 as usual denigrated the South African Government accusing it of being a 'racist régime' repressing the profound aspirations of peoples. But it also denigrated the member nations of the North Atlantic Treaty Organization, specifically France, the United Kingdom and the United States of America, accusing them of aiding and abetting such repression or, worse still, strongly condemning them.

Can these Western permanent members of the Security Council really put up with such actions on the part of the General Assembly?

This question was answered by the nine countries of the European Community on 26 January 1976, when on the eve of the new debate in the Security Council on the future of Namibia they addressed themselves to the Government of South Africa under advice to the United Nations (S/11945—see appendix 6, page 98). In their letter the nine countries felt that it is essential for the Republic of South Africa to promote the process of self-determination of Namibia on the basis of a number of enumerated principles and by applying modalities which should be termined through negotiations between the South African Government and the Secretary-General of the United Nations so as to enable the United Nations to exercise supervision.

The nine nations thereby dissociated themselves from the demand that South Africa should withdraw its administration from Namibia prior to the completion of the process of self-determination. Neither did the resolution of 30 January 1976, with which the Security Council closed its debate, demand an earlier withdrawal of South Africa, nor were any decisions taken in respect of sanctions.

The spokesman for South Africa, Mr. R. F. Botha could fully and without hindrance develop the case for his country in the Security Council's session (S/PV 1881 pages 37-58; S/11948 and Addendum Nr. 1).

The Security Council did not create any room for the UN Council for Namibia to initiate any activities within the territory prior to the moment of realisation of the territory's self-determination. The French representative Mr. de Guiringaud drew attention to this point on 30 January 1976 after the decision on the draft resolution was made. The role of the United Nations in the process of self-determination should be both effective and precise: 'It seems to us, however'—said the spokesman for France—'that the supervision should not result in the creation of a machinery incompatible with the presence of an administration which will not have left the Territory before the elections are held.' (S/PV.1885). The UN Council for Namibia and the United Nations Commissioner for Namibia could just as well shelve their plans to create such a machinery and forget about them. And so it was intended by the nine countries of the European Community.

# XVI

# South Africa's response to the UN over the South West Africa issue

The preceding chapters set out in broad outline UN—South African relations over the South West Africa issue. Little has been said of South Africa's point of view. Although the scope is necessarily limited, this chapter is intended to fill that gap. As far as possible, official South African spokesmen are quoted.

South Africa has repeatedly stated that it has no claims to South West Africa or to any part of its territory. Apart from official statements of policy, this attitude reflected also in the reports of the discussions which took place between the Secretary-General and the Prime Minister, and also in the reports of the discussions which Dr. Escher had with South African officials. In the letter the South African Minister of Foreign Affairs, Dr. H. Muller, addressed to Dr. Waldheim on 27 May 1975, Dr. Muller wrote:

> My government has repeatedly stated that it recognises the distinct international status of South West Africa and that it does not claim one inch of the territory for itself. Its sole concern has been to develop the territory in the best interests of all its inhabitants and to prepare them for the orderly exercise of their right of self-determination.

South Africa has also repeatedly stated that its goal for South West Africa is self-determination and independence. This much was confirmed by the Secretary-General following his meeting with the South African Prime Minister, Mr. B.J. Vorster, in March 1972. In his report Dr. Waldheim declared: 'With regard to the question of Namibia, the South African government confirms that its policy is one of self-determination and independence' (S/10738). And the South African Minister of Foreign Affairs, in his statement in the general debate in the General Assembly of 5 October 1973, spoke as follows:

> South Africa's presence in the territory is accounted for by the undertaking we have given to ensure that its inhabitants are able and ready to exercise their rights of self-determination and independence. We have no design whatsoever with respect to any part of South West Africa's territory. In the meanwhile, we believe we have a duty to continue to assist the development of South West Africa in all spheres: economic, social and political.

South Africa's contribution to the social and economic development of South West Africa is impressive. Thus an investment corporation for South West Africa has formulated a programme which will create 5.000 jobs for blacks during the period 1972-1977, entailing a capital investment in excess of R22 million.* An amount of R139 million has been spent upto 1974 on 177 domestic water supply schemes constructed and operated by the state throughout the territory. Educational opportunities for persons of colour have been vastly expanded. Thus the number of schools has increased from 313 in 1960 to 592 in 1973; the number of teachers from 1.310 in 1960 to 3.453 in 1973; and the number of pupils from 43.000 in 1960 to 140.000 in 1973.

With regard to the transportation system, the total investment of the South African Railways Administration in the territory amounted in 1973 to R170 million; and the total outlay on roads from 1953 to 1973 exceeded R243 million. The value of telephones, telegraph and radio installations in South West Africa amounted to R35 million in 1973. And the total cost of administering the territory now amounts to R341 million per annum. This is a considerable figure, bearing in mind that the present total population is only 850.000.

Impressive as the socio-economic features of South Africa's stewardship of South West Africa unquestionable are, it is the political aspect which interests most of South Africa's critics. What is the position here?

The point has already been made that South Africa has no territorial ambitions in regard to South West Africa, and, moreover, that its objective is independence and self-determination. It is on this aspect that differences of opinion occur. Most of South Africa's critics at the UN understand self-determination to mean self-government on majority rule basis, regardless of the great demographic and cultural diversity of the population. They advocate what we have termed 'fictitious self-determination'. By contrast, the South African government has stressed the heterogenety of the population and the

* R 1 equals about US $ 1.20

fact that the country is composed of many peoples, each of which has a right to maintain its cultural identity. However, the South African government has also consistently stressed that the political future of South West Africa must be determined by the people themselves.

The South African Minister of Foreign Affairs, in his letter of 27 May 1975, to the Secretary-General, expressed this point as follows:

> As is well-known, the basis of the South African government's approach to the question of South West Africa is that it is for the people of South West Africa themselves to determine their own political and constitutional future in accordance with their own freely expressed wishes. This presupposes that they should exercise their choice freely and without interference from South Africa, the United Nations, or any other outside entity. All options are therefore open to them—including that of independence as one state, if that is what they should choose.

In accordance with a procedure formulated in consequense of the above, representatives of all population groups in South West Africa participated in the earlier mentioned constitutional conference from 1 to 12 September 1975, and are continuing their work. In a major policy statement on South West Africa, the Prime Minister, in a speech in Windhoek on 20 May 1975, said of this process of 'mutual consultation': 'Nobody, surely, is under the illusion that it will be an easy process, but the prospect does exist that agreement can be reached among all population groups'. Except to express his conviction '. . . that the peoples of South West Africa will not be so shortsighted as to pursue a course which will disrupt the foundations of the economic system of the territory, to the material detriment of all the population groups which have their homes there', Mr. Vorster emphasized that, as far as the South African government is concerned, all constitutional options are open to the people of South West Africa.

The *bona fides* of the South African government, in this respect and others, may be judged from the following remarks of the Prime Minister during the course of the same speech:

> In substance we are in agreement with the most important aspects of the points of view which are put in the United Nations. As far as the O.A.U. are concerned, in principle, and bearing in mind what I have already said, we have no quarrel with their points of view concerning self-determination, independence and the maintenance of the territorial integrity of the Territory. Where we do differ and differ

vere clearly, is in regard to the role claimed for the United Nations and SWAPO. We do not hide, nor have we ever been ashamed of, our administration of the Territory. Indeed, we go out of our way to make information concerning the Territory and its people freely available.

I also want to repeat here what I have said in the House of Assembly, viz., that leaders of Africa who may be interested in visiting the Territory in order to acquaint themselves with conditions there, are very welcome to do so, either personally or through their representatives.

I would also be prepared, as I have said before, to exchange ideas with a committee of the O.A.U. on the basis of the points of view which I have expounded here, but as I have already stated in the clearest possible terms, I cannot accept the rôle which is proposed for SWAPO.

If the African Chairman of the Council for South West Africa, and the Special Committee of the O.A.U., are therefore really interested in discussing with me in South Africa the progress of the peoples of the Territory, I shall also ask the real leaders of the peoples of the Territory to talk to them. Furthermore, if they would like to receive these leaders in their own countries in order to obtain from them first hand information, they are free to invite them, and we shall do all we can to help make the visit or visits possible.

# XVII

# South Africa and Africa

The difficulties which could lie ahead in South Africa's relations with the UN are not likely to be resolved in New York but in Africa itself. For with Portugal's decision to withdraw from its African colonies as the catalyst, South Africa appears to enter a new and important phase in its relations with other African countries.

South Africa's relations with the rest of Africa in the post-colonial period have been strained, because as countries to South Africa's north became independent, at the UN and elsewhere, they increasingly set the tone of international criticism of South Africa. As a matter of fact, this situation led many outside observers to conclude in the late' fifties that confrontation and conflict between South Africa and the rest of the continent was inevitable. That the confrontation did not occur was due to several complex factors, among which, however, was unquestionably South Africa's economic strength relative to most other African countries, its relative political stability, and the fact that the official race relations policy continued to evolve in the face of both internal and external demands for change.

For most of the 'fifties and also the 'sixties, South Africa adopted a highly defensive foreign policy posture. Thus at the UN it relied heavily on the domestic jurisdiction clause (Art. 2-7 Charter)—refusing even to discuss its internal policies—and in its attitude towards other African countries, it assumed a 'low profile'. A change was evident in the second half of the 'sixties when, with the so-called 'outward policy', South Africa felt that the international environment had changed sufficiently for it to adopt a more enterprising foreign policy. The Government also felt that its internal policies had evolved to the point where it had something to show.

As a part of this initiative South Africa launched a policy of dialogue in Africa which, over the period 1967 to 1969, looked promising as several, primarily francophone, countries expressed an interest in contact. However, this movement appeared to have been premature. It was also negatively effected by events in Rhodesia at the time and the failure to reach a constitutional settlement. With the exception of Malawi, South Africa's efforts at establishing diplomatic ties were fruitless.

The Portuguese decision to withdraw from the African territories was earlier said to have had a catalytic effect on relations in South Africa. In the interim period between the publication of the Lusaka Manifesto of April 1969 (which is a convenient marker of the end of the first dialogue movement), and the coup in Lisbon of April 1974, South Africa gave little public indication of its intentions with regard to Africa. But that a certain amount of behind-the-scenes exploratory diplomacy was occuring seems very likely. However, whatever the facts following the coup in Lisbon the South African government felt obliged to renew its efforts at dialogue. This was first heralded in a widely acclaimed interview which the South African Prime Minister gave to Newsweek in September 1974, in which he emphasized that it was not important to South Africa who ruled in Mozambique and Angola, but that there should be political stability in these countries and in the region as a whole. This, said the Prime Minister, was South Africa's primary concern.

Mr. Vorster followed this up with a major speech in Parliament on 23 October 1974. His subject was foreign relations in Southern Africa, and the Prime Minister's main theme was that the African countries faced a choice. In his own words –

> I believe that South Africa has come to the crossroads. I believe that Southern Africa has to make a choice. I think that choice lies between peace on one hand and escalation of strife on the other. The consequences of an escalation are easily foreseeable. The toll of a major confrontation will be high. I would go so far as to say that it will be too high for Southern Africa to pay. If one adds to that the threatening economic problems which could assume major proportions, then Africa and Southern Africa should guard against heading for chaos.
>
> However, this is not necessary . . . for there is an alternative way. That way is the way of peace. The way of normalizing of relations, the way of sound understanding and normal associations. I believe that Southern Africa can take that way. I believe that it is prepared to prefer to take that way, and I believe that it will do so in the end.

Within a matter of days, the South African Prime Minister's statement was publicly welcomed by the President of Zambia, Dr. Kenneth Kaunda, who described it as the 'voice of reason'. Various other developments followed: the South African Prime Minister visited Liberia and Ivory Coast for talks with his counterparts there; South Africa, in apparent co-operation with several other states in the Southern African region, encouraged renewed efforts at reaching a constitutional settlement of the Rhodesian question; and South Africa took some bold initiatives in seeking a solution to the South West African issue.

These encouraging developments led observers to describe the Southern African situation as one of détente. And in 1975, South Africa may be said to be moving cautiously towards a relationship of peaceful co-existence with its neighbours. This is what South Africa obviously wants. But equally important is what other countries in Southern Africa would wish to happen. And this is bound also profoundly to influence attitudes towards South Africa at the UN, the other countries being greatly influenced by the feelings of the UN.

It would be appropriate for the UN in any case to foster peaceful co-existence in Southern Africa. The General Assembly should therefore refrain from taking decisions and issuing statements which could damage the feelings of mutual appreciation and respect of the nations of Southern Africa, as these are an important condition for a succesful peaceful co-existence. Likewise the UN should cease to conduct and support hostile propaganda against South Africa. The Security Council, ultimately responsible for solving situations arising out of failures to co-exist peacefully, should reasonably be able to expect this from the General Assembly.

In 1975 however this position was no where reached. During the 30th session of the General Assembly and—so it should be noted—with South Africa *in absentia,* decisions were taken which could only create and stimulate disrespect and enmity towards South Africa amongst the other nations of Southern Africa.

It is now upon the UN Members not wishing to accept this misapplication of the Charter, and most certainly upon the Western permanent members of the Security Council who will not accept misapplication lest it would affect the position of the Security Council and thereby their own position as permanent members, to consider how to assist South Africa in its aim to reach a relationship of peaceful co-existence with its neighbours.

# XVIII

# Whither, United Nations?

The maintenance of international peace and security in Southern Africa depends in large measure upon the success which the process of implementation of the principle of self-determination of peoples—now growing in South Africa, South West Africa (Namibia), Mozambique and Angola, and about to commence in Rhodesia—will have.

If such a process deteriorates into war, as has been the case in Angola, a situation arises which can be termed a real danger to international peace and security and which therefore is ripe for investigation by the Security Council according to Article 34 of the Charter. If South Africa should abandon the peoples of South West Africa, leaving the territory open to invasion by SWAPO, a situation similar to the one which has arisen in Angola may develop.

The political development in all these territories should seriously trouble the UN. In this respect it cannot but be noticed that South Africa through its policy of détente is doing all within its power to alleviate those troubles for the UN. Among the principal statesmen in Southern Africa who clearly appreciate this, is President Kaunda of Zambia who, together with Prime Minister Vorster of South Africa, is now acting as an intermediary in the Rhodesian situation.

Under such conditions the Security Council cannot very well continue to accept that South Africa's ability succesfully to carry out its policy of détente should be thwarted by activities on the part of the General Assembly which undermine the authority and credibility of South Africa in the world. It would be in the interest of the UN Organization to undo not only the General Assembly's decision to debar South Africa from its work, but also those resolutions which in violation of the Charter interfere with the sovereignty of South Africa (see page 59 seq.).

In case it should prove impossible to persuade the General Assembly to revoke those defective decisions some other way must be found to neutralize them. Perhaps they could be allowed quietly to expire. In any case they should not be allowed to be confirmed as has been done during the past years. If however the majority of the General Assembly under the influence of the Special Committee against apartheid persists in its efforts to eliminate South Africa, developments cannot possibly be predicted.

The UN Organization occupies a central position in the sphere of international relations and therefore determines to a considerable extent the nature of the interrelationships between the member nations.

The elimination of a Member contrary to the provisions of the Charter renders it necessary for those Members who cannot accept its validity to establish a new legal relationship with the victim-Member in lieu of its former relationship through the UN.

During the debate in the General Assembly from 6 to 12 December 1974 on the subject 'Strengthening of the role of the United Nations with regard to the maintenance and consolidation of international peace and security, the development of co-operation among all nations and the promotion of the rules of international law in relations between States', spokesmen of the great Western Powers voiced their objections to the elimination of South Africa from the General Assembly which constituted a circumvention or infringement of the Charter, making it clear that they would not resign themselves to the situation.

The majority of the nations taking part in the debate did not appear fully to appreciate the true nature of these objections. Indications are that this majority will continue to use the UN Organization to settle their accounts with and to deal with South Africa, and in due course also with Israel, and then with other Members out of favour with the majority.

It looks as if the Western Powers will not be able to forestall these efforts by mere discussions in the UN meetings. They will need to persuade and convince the other member nations about the meaning and validity of the Charter as Treaty indicating what consequences they will draw for themselves from infringements of the Treaty which may effect their membership and the discharge of their obligations which arise from their membership.

One may expect that these subjects will be extensively discussed outside the precincts of the UN. The practical feasibility and the life expectancy of the UN may be said to be dependent upon the results of these discussions. They could be not only essential and indispensable, but also useful for 'the development of co-operation among all nations and the promotion of the rules of international law.' They could be a training course for the revalidation of the interrelationships of UN Members.

Denigration, abusive language, false or exaggerated reproaches, insufficient preparedness to listen to each other, all this could make the negotiations *extra muros* senseless and should therefore be avoided on this level. Only then could chances for true consultation and deliberation and appeacement be used, chances which in the workings of the UN have been somewhat lost in the past. Only then would the issues of South Africa and South West Africa be judged on their merits.

If the result could be achieved that South Africa and the other countries in Southern Africa be involved in such negotiations, it might be possible that these countries would turn to a peaceful co-existence, which would leave no room for complaints at the UN, no room to conduct hostile propaganda against each other, nor for the encouragement of terrorism.

In short this might be the most real contribution to the 'maintenance and consolidation of international peace and security' in Southern Africa, a state of affairs which as we hope may one day find its consolidation by the United Nations.

Appendix 1

# Original text letter King William III of the Netherlands of 14 July, 1855*

*Brief van Z. Majesteit den Koning van Holland aan Z.H. Ed. den President van den Oranje Vrystaat*

Mynheer de President, Omstandigheden, waaraan myne voorvaderen even onschuldig waren als de Uwen, hebben de gelukkige bevolking en de bloeijende landstreken, waarover Gy thans het bewind voert, aan hun gezag onttrokken. Diep is zulks dezerzyds betreurd, omdat geheel Nederland innig aan die bevolking gehecht was, gelyk zy wederkeerig aan Oranje en Nederland. De band dier wederkeerige gehechtheid bleef intusschen bestaan, ofschoon de staatkundige banden verbroken waren. Treffende blyken zyn daarvan herhaaldelyk van Uwe zyde en die Uwer landgenoten geleverd. De trouw van geslotene traktaten en van eene eerlyke Staatkunde verbood ons echter eenig dergelyk bewys te geven, hoezeer ook het Vaderlandsch gemoed daartoe gezind zoude zyn geweest.

Sedert is de onafhankelijkheid en zelfstandigheid van den Oranje-Vrystaat op wettige wijze erkend geworden; van dit oogenblik stond het u vry, en was het ook ons geoorloofd, lucht te geven aan de gevoelens van warme verkleefdheid, door aloude herinneringen en gemeenschappelyken oorsprong diep in het harte geplant, en nimmer verzwakt.

By officiëlen brief van 15 October 1854, hebt Gy, Mynheer de President, My kennis gegeven, dat by wettig verdrag, de Oranje-Vrystaat tot een' onafhankelyken en vryen staat is verklaard.

Ontvang, met myne opregte gelukwenschen mynen dank voor die mededeeling. Zy vervulde my met groot blydschap. God, die myne en Uwe, onze gemeenschappelyke vaderen, in zoo menigen bangen stryd voor vryheid en onafhankelykheid, zoo genadiglyk beschermde en krachtig ondersteunde, neme de Oranje-Vrystaat, by voortduring in Zyne alvermogende hoede!

Gelyk vroeger myn Doorluchtige Voorvader, Willem de Zwijger, en Zyne opvolgers, met hunne getrouwe Nederlanders, voor de heilige zaak van Vryheid, Verlichting, en Beschaving streden, zyt Gylieden, als een moedige voorpost, steeds strydende

* See page 4, seq.

tegen de barbaarschheid en het ongelyk, welke U gestadig dreigen te overweldigen. Moge Uwe stryd duurzaam zegevierend en zoo min mogelyk verwoestend, voor Uwe welvaart zyn! Moge de tak van den Oud-Hollandschen Stam, welke by Ulieden wortel heeft geschoten, en U volgens Uwe eigene uitdrukking, tot eene andere Hollandsche Natie heeft doen opwassen, krachtvol voortgroeijen, en Uwe roem en welvaart, welke ik met welgevallen en dankbaarheid zie dat by aan den naam van myn stamhuis hebt vastgeknoopt, steeds vermeerderen!

Ik zeg U, Mynheer de President, dank voor al de betuigingen van verkleefdheid, welke ik in Uwen brief mogt aantreffen, en verzoek U, overtuigd te zyn, dat die gevoelens by My, zoowel als by myne beminde onderdanen, den warmsten weerklank vinden.

Gaarne wensch ik U daarvan al die bewyzen te geven, welke in myne magt staan.

Het was my daarom hoogst aangenaam den Oranje-Vrystaat al dadelyk een bewys van zyne erkenning te kunnen geven, door myn exequatur verleenen aan den door U hier te lande benoemden Consul.

Door hem is, namens Ulieden verlangd, dat Vlag en Wapen door My aan Uwen Staat zouden worden geschonken.

Voor zooverre zulks uw voortdurende wensch blyft, zal het my zeer welkom zyn, tot herinnering aan de naauwe banden, welke tusschen Onze voorvaderen en de Uwen bestonden, en tot bevestiging van die, welke Wy wenschen dat tusschen Ons en U blyven bestaan, aan dat verlangen te voldoen.

Hy, die deze Mynen brief aan U zal ter hand stellen, is belast u meer breedvoerig over dit onderwerp te onderhouden.

Ik bid God, Mynheer de President, dat Hy U en de Uwen in Zyne heilige bescherming gelieve te nemen, en gaarne neem ik deze gelegenheid te baat, om U de verzekering te geven der gevoelens van hoogachting, waarmede ik my noeme, Mynheer de President, Uwen goeden Vriend.

WILLEM

Het Loo, den 14den July, 1855.

Appendix 2

# G.A. Declaration on non-intervention*

*2131 (XX). Declaration on the Inadmissibility of Intervention in the Domestic Affairs of States and the Protection of their Independence and Sovereignty*

*The General Assembly,*

*Deeply concerned* at the gravity of the international situation and the increasing threat to universal peace due to armed intervention and other direct or indirect forms of interference threatening the sovereign personality and the political independence of States,

*Considering* that the United Nations, in accordance with their aim to eliminate war, threats to the peace and acts of aggression, created an Organization, based on the sovereign equality of States, whose friendly relations would be based on respect for the principle of equal rights and self-determination of peoples and on the obligation of its Members to refrain from the threat or use of force against the territorial integrity or political independence of any State,

*Recognizing* that, in fulfilment of the principle of self-determination, the General Assembly, in the Declaration on the Granting of Independence of Colonial Countries and Peoples contained in resolution 1514 (XV) of 14 December 1960, stated its conviction that all peoples have an inalienable right to complete freedom, the exercise of their sovereignty and the integrity of their national territory, and that, by virtue of that right, they freely determine their political status and freely pursue their economic, social and cultural development,

*Recalling* that in the Universal Declaration of Human Rights the General Assembly proclaimed that recognition of the inherent dignity and of the equal and inalienable rights of all members of the human family is the foundation of freedom, justice and peace in the world, without distinction of any kind,

* See page 48-50, 61.

*Reaffirming* the principle of non-intervention, proclaimed in the charters of the Organization of American States, the League of Arab States and the Organization of African Unity and affirmed at the conferences held at Montevideo, Buenos Aires, Chapultepec and Bogotá, as well as in the decisions of the Asian-African Conference at Bandung, the First Conference of Heads of State or Government of Non-Aligned Countries at Belgrade, in the Programme for Peace and international Co-operation adopted at the end of the Second Conference of Heads of State or Government of Non-Aligned Countries at Cairo, and in de declaration on subversion adopted at Accra by the Heads of State and Government of the African States,

*Recognizing* that full observance of the principle of the non-intervention of States in the internal and external affairs of other States is essential to the fulfilment of the purposes and principles of the United Nations,

*Considering* that armed intervention is synonymous with aggression and, as such, is contrary to the basic principles on which peaceful international co-operation between States should be built,

*Considering further* that direct intervention, subversion and all forms of indirect intervention are contrary to these principles and, consequently, constitute a violation of the Charter of the United Nations,

*Mindful* that violation of the principle of non-intervention poses a threat to the independence, freedom and normal political, economic, social and cultural development of countries, particularly those which have freed themselves from colonialism, and can pose a serious threat to the maintenance of peace,

*Fully aware* of the imperative need to create appropriate conditions which would enable all States, and in particular the developing countries, to choose without duress or coercion their own political, economic and social institutions,

*In the light of the foregoing considerations, solemnly declares:*

1. No State has the right to intervene, directly or indirectly, for any reason whatever, in the internal or external affairs of any other State. Consequently, armed intervention and all other forms of interference or attempted threats against the personality of the State or against its political, economic and cultural elements, are condemned.

2. No State may use or encourage the use of economic, political or any other type of measures to coerce another State in order to obtain from it the subordination of the exercise of its sovereign rights or to secure from it advantages of any kind. Also, no State shall organize, assist, foment, finance, incite or tolerate subversive, terrorist or armed activities directed towards the violent overthrow of the régime of another State, or interfere in civil strife in another State.

3. The use of force to deprive peoples of their national identity constitutes a violation of their inalienable rights and of the principle of non-intervention.

4. The strict observance of these obligations is an essential condition to ensure that nations live together in peace with one another, since the practice of any form of intervention not only violates the spirit and letter of the Charter of the United Nations but also leads to the creation of situations which threaten international peace and security.

5. Every State has an inalienable right to choose its political, economic, social and cultural systems, without interference in any form by another State.

6. All States shall respect the right of self-determination and independence of peoples and nations, to be freely exercised without any foreign pressure, and with absolute respect for human rights and fundamental freedoms. Consequently, all States shall contribute to the complete climination of racial discrimination and colonialism in all its forms and manifestations.

7. For the purpose of the present Declaration, the term "State" covers both individual States and groups of States.

8. Nothing in this Declaration shall be construed as affecting in any manner the relevant provisions of the Charter of the United Nations relating to the maintenance of international peace and security, in particular those contained in Chapters VI, VII and VIII.

*1408th plenary meeting,*
*21 December 1965.*

## Appendix 3

# Draft resolution rejected by the Security Council in terms of article 27 UN Charter*

S/11543—24 October, 1974

*Kenya, Mauritania and United Republic of Cameroon: draft resolution*

*The Security Council,*

*Having considered* General Assembly resolution 3207 (XXIX) of 30 September 1974, in which the Assembly called upon the Security Council 'to review the relationship between the United Nations and South Africa in the light of the constant violation by South Africa of the principles of the Charter and the Universal Declaration of Human Rights',

*Having heard* the statements of the persons invited to address the Council on this issue,

*Taking note* of the special report of the Special Committee on *Apartheid* on 'violations of the Charter of the United Nations and resolutions of the General Assembly and the Security Council by the South African régime' (S/11537),

*Mindful* of the provisions of the Chartter of the United Nations concerning the rights and obligations of Member States, particularly Articles 1, 2, 6, 55 and 56,

*Recalling* its resolutions 134 (1960), 181 (1963), 182 (1963), 190 (1964), 282 (1970), and 311 (1972) on the question of the policies of *apartheid* of the Government of the Republic of South Africa,

*Reaffirming* that the policies of *apartheid* are contrary to the principles and purposes of the Charter of the United Nations and inconsistent with the provisions of the Universal Declaration of Human Rights, as well as South Africa's obligations under the Charter,

* See page 63, seq.

*Recalling* that the General Assembly and the Security Council have more than once condemned the South African Government for its persistent refusal to abandon its policies of *apartheid* and to abide by its obligations under the Charter, as called for by the Security Council and the General Assembly,

*Noting with concern* South Africa's refusal to withdraw its police and military forces, as well as its civilian personnel, from the mandated Territory of Namibia and to co-operate with the United Nations in enabling the people of Namibia as a whole to attain self-determination and independence,

*Noting further* that, in violation of the pertinent resolutions of the Security Council, particularly resolution 253 (1968) of 29 May 1968, South Africa has not only given support to the illegal régime in Southern Rhodesia, but has also sent into that Territory military and police personnel for the purpose of strengthening that régime in its attemtp to impede the exercise of their inalienable rights by the people of that Territory,

*Considering* that effective measures should be taken to resolve the present situation arising out of the policies of *apartheid* of the Government of South Africa,

*Recommends* to the General Assembly the immediate expulsion of South Africa from the United Nations in compliance with Article 6 of the Charter.

## Appendix 4

# Draft resolution rejected by the Security Council in terms of article 27 UN Charter*

S/11713—6 June, 1975

*Guyana, Iraq, Mauritania, the United Republic of Cameroon and the United Republic of Tanzania: draft resolution*

*The Security Council,*

*Recalling* General Assembly resolution 2145 (XXI) of 27 October 1966, which terminated South Africa's mandate over the Territory of Namibia, and resolution 2248 (S-V), of 19 May 1967, which established the United Nations Council for Namibia, as well as all other subsequent resolutions on Namibia, in particular resolution 3295 (XXIX) of 13 December 1974,

*Recalling* Security Council resolution 245 (1968) of 25 January and 246 (1968) of 15 March 1968, 264 (1969) of 12 August 1969, 276 (1970) of 30 January, 282 (1970) of 23 July, 283 (1970) and 284 (1970) of 29 July 1970, 300 (1971) of 12 October and 301 (1971) of 20 October 1971, 310 (1972) of 4 February 1972 and 366 (1974) of 17 December 1974, which confirmed General Assembly decisions.

*Recalling* the advisory opinion of the International Court of Justice of 21 June 1971 that South Africa is under obligation to withdraw its presence from the Territory,

*Taking note* of the letter dated 27 May 1975, addressed by the Minister for Foreign Affairs of South Africa to the Secretary-General (S/11701),

*Having heard* the statement by the President of the United Nations Council for Namibia,

*Having considered* the statement by Mr. Sam Nujoma, President of the South-West Africa People's Organization (SWAPO),

* See page 70, seq.

*Gravely concerned* about South Africa's continued illegal occupation of Namibia and its persistent refusal to comply with resolutions and decisions of the General Assembly and the Security Council, as well as with the advisory opinion of the International Court of Justice of 21 June 1971,

*Gravely concerned* at South Africa's brutal repression of the Namibian people and its persistent violations of their human rights, as well as its efforts to destroy the national unity and territorial integrity of Namibia,

*Reaffirming* the inalienable and imprescriptible rights of the people of Namibia to self-determination, national independence and the preservation of their territorial integrity,

*Noting with concern* that South Africa has not made the declaration demanded in paragraph 3 of resolution 366 (1974) of the Security Council,

*Further noting* with the deepest concern that the demands in paragraphs 4 and 5 in the aforementioned resolution have been totally ignored by South Africa,

1. *Condemns* South Africa's failure to comply with terms of Security Council resolution 366 (1974) of 17 December 1974;

2. *Condemns once again* the continued illegal occupation of the Territoyr of Namibia by South Africa;

3. *Further condemns* the illegal and arbitrary application by South Africa of racially discriminatory and repressive laws and practices in Namibia;

4. *Demands* that South Africa put an end forthwith to its policy of bantustans and the so-called homelands aimed at violating the national unity and the territorial integrity of Namibia;

5. *Further demands* that South Africa proceed urgently with the necessary steps to withdraw from Namibia and, to that end, to implement the measures stipulated in resolution 366 (1974);

6. *Reaffirms* the legal responsibility of the United Nations over Namibia and demands that South Africa take appropriate measures to enable the United Nations Council for Namibia to establish its presence in the Territory with a view to facilitating the transfer of power to the people of Namibia;

7. *Declares* that in order for the people of Namibia to freely determine their own future it is imperative that free elections be organized under the supervision and control of the United Nations as soon as possible and, in any case, not later than 1 July 1976;

8. Affirms its support for the struggle of the people of Namibia for self-determination and independence;

9. *Acting* under Chapter VII of the United Nations Charter,

*a. Determines* that the illegal occupation of the Territory of Namibia by South Africa constitutes a threat to international peace and security;

*b. Decides* that all States shall prevent:

(i) Any supply of arms and ammunition of South Africa;

(ii) Any supply of aircraft, vehicles and military equipment for use of the armed forces and paramilitary organizations of South Africa;

(iii) Any supply of spare parts for arms, vehicles and military equipment used by the armed forces and paramilitary organization of South Africa;

(iv) Any activities in their territories which promote or are calculated to promote the supply of arms, ammunition, military aircraft and military vehicles to South Africa and equipment and materials for the manufacture and maintenance of arms and ammunition in South Africa and Namibia;

10. *Decides* that all States shall give effect to the decision set out in paragraph 9 (b) of this resolution notwithstanding any contract entered into or licence granted before the date of this resolution, and that they notify the Secretary-General of the measures they have taken to comply with the aforementioned provision;

11. *Decides* that provisions of paragraph 9 (b) shall remain in effect until it has been established, to the satisfaction of the Security Council, that the illegal occupation of the Territory of Namibia by South Africa has been brought to an end;

12. *Requests* the Secretary-General, for the purpose of the effective implementation of this resolution, to arrange for the collection and systematic study of all available data concerning international trade in the items which should not be supplied to South Africa under paragraph 9 (b) above;

13. *Requests* the Secretary-General to report to the Security Council concerning the implementation of paragraph 7 and other provisions of this resolution;

14. *Decides* to remain seized of the matter and to meet on or before 30 September 1975 for the purpose of reviewing South Africa's compliance with the terms of the relevant paragraphs of this resolution, and in the event of non-compliance by South Africa to taking further appropriate measures under the Charter.

Appendix 5

# Document of the South West Africa Conference of September 1975*

The South West Africa Constitutional Conference commenced with its first session on 1 September 1975 and at the conclusion of that phase of the Conference, a Declaration of Intent regarding the future constitutional development of the Territory was adopted. Te Declaration reads as follows:

We, the true and authentic representatives of the inhabitants of South West Africa, hereby solemnly declare;

That in the exercise of our rights to self-determination and independence, we are voluntarily gathered in this Conference in order to discuss the constitutional future of South West Africa;

That we most strongly condemn and reject the use of force or any improper interference in order to overthrow the existing order or to enforce a new dispensation;

That we are firmly resolved to determine our future ourselves by peaceful negotiation and co-operation;

That mindful of the particular circumstances of each of the population groups it is our firm resolve in the execution of our task to serve and respect their whishes and interests;

That the mindful of the interdependence of the various population groups and the interests of South West Africa in its entirety, we are therefore resolved to create a form of Government which wil guarantee to every population group the greatest possible say in its own and national affairs, which will fully project the rights of minorities and which will do justice to all;

And further we declare that we are resolved to devote continuous attention to social and economic conditions which will best promote the welfare, interests and peaceful coexistence of all the inhabitants of South West Africa and their descendants;

That we are resolved to devote ourselves to the promotion of and respect for human rights and fundamental freedoms for all without discrimination merely on the basis of race, colour or creed;

We therefore decide:

* See chapter XV.

(a) To draft a constitution for South West Africa as soon as possible and, if possible within a period of three years;

(b) To devote continuous attention to measures to implement all the aims specified in this Declaration.'

The second session of the Conference took place from 10 to 14 November 1975. The Conference decided that the agenda items should be referred to committees for in-depth study and four committees, known as the First, Second, Third and Fourth Committees were established, the first two having 11 representatives each and the other two, 10 members. Each was authorized to hear expert evidence concerning its particular field of activity, as and when considered necessary. The Conference requested the Committees to commence work immediately following the conclusion of the second session of the Conference and to report to the full Conference when the third session convenes early in March 1976.

## Appendix 6

# Text of demarche of the Nine of the EEC

S/11945—27 January, 1976

*Letter dated 26 January 1976 from the permanent representative of the Netherlands to the United Nations addressed to the Secretary-General**

I have been requested by the Government of the Grand Duchy of Luxembourg, which country currently exercises the function of President of the European Community, to transmit to you the outlines of the *démarche* undertaken on the eve of the debate in the Security Council on Namibia by the Netherlands Ambassador to South Africa on behalf of the nine countries of the Community.

The nine countries of the European Community have had an exchange of views on the problem of Namibia within the framework of their normal consultations on political co-operation. With reference to what has emerged from the Windhoek constitutional conference and from the debate which took place recently in the United Nations General Assembly, they deem it advisable to set forth to the Government of the Republic of South Africa their position on the question of Namibia.

As they indicated in the message addressed to the Secretary-General of the United Nations on the occasion of 'Namibia Day', the nine countries of the European Community consider that South Africa should withdraw from Namibia as soon as possible and that the inhabitants of Namibia should have the opportunity to exercise at an early date their right to self-determination and to independence. The exercise of this right should consist of the opportunity of expressing their opinions on the political and constitutional future of the Territory as a whole, through a fully democratic process under the supervision of the United Nations.

They accordingly feel that all the political groups should be free to engage in peaceful political activities throughout the Territory during the process of self-determination.

Consequently, they consider that the release of all Namibians detained because of their political opinions and the return to their territory of the Namibians at present in exile could contribute appreciably to ensuring this democratic and peaceful participation in the process of self-determination.

In the light of the foregoing, the nine countries consider the indications that have emerged thus far from the Windhoek constitutional conference inadequate. As its

* See page 75.

representativity is limited to ethnic groups, not all the political forces have been able to participate in it. It therefore does not appear to guarantee the fully democratic character of the process of self-determination. Furthermore, it still does not appear to provide for the constitutional future of Namibia to be determined by the Namibian people through a single consultation organized on a Territory-wide basis.

The nine countries of the Community therefore feel that it is essential for the Republic of South Africa to promote the process of self-determination of Namibia on the basis of the above-mentioned principles and by applying modalities which should be determined through negotiations between the South African Government and the Secretary-General of the United Nations so as to enable the United Nations to exercise supervision.'

I have the honour to request you to transmit this letter to the President of the Security Council and to have it circulated as an official document of the Security Council.

(*Signed*) Johan Kaufmann
Permanent Representative of the
Kingdom of the Netherlands
to the United Nations

# Appendix 7

# Text United Nations Charter

## THE CHARTER OF THE UNITED NATIONS

*We the Peoples of the United Nations*

*Determined*

to save succeeding generations from the scourge of war, which twice in our lifetime has brought untold sorrow to mankind, and

to reaffirm faith in fundamental human rights, in the dignity and worth of the human person, in the equal rights of men and women and of nations large and small, and

to establish condtions under which justice and respect for the obligations arising from treaties and other sources of international law can be maintained, and

to promote social progress and better standards of life in larger freedom,

*And for these ends*

to practice toerance and live together in peace with one another as good neighbours, and

to unite our strength to maintain international peace and security, and

to ensure by the acceptance of principles and the institution of methods, that armed force shall not be used, save in the common interest, and

to employ international machinery for the promotion of the economic and social advancement of all peoples,

*Have resolved to combine our efforts to accomplish these aims*

Accordingly, our respective Governments, through representatives assembled in the city of San Francisco, who have exhibited their full powers found to be in good and due form, have agreed to the present Charter of the United Nations and do hereby establish an international organization to be known as the United Nations.

APPENDICES

## I. PURPOSES AND PRINCIPLES

**Art. 1.** The Purposes of the United Nations are:

1. To maintain international peace and security, and to that end: to take effective collective measures for the prevention and removal of threat to the peace, and for the suppression of acts of aggression or other breaches of the peace, and to bring about by peaceful means, and in conformity with the principles of justice and international law, adjustment or settlement of international disputes or situations which might lead to a breach of the peace;
2. To develop friendly relations among nations based on respect for the principle of equal rights and self-determination of peoples, and to take other appropriate measures to strengthen universal peace;
3. To achieve international so-operation in solving international problems of an economic, social, cultural, or humanitarian character, and in promoting and encouraging respect for human rights and for fundamental freedoms or all without distinction as to race, sex language, or religion; and
4. To be a center for harmonizing the actions of nations in the attainment of these common ends.

**Art. 2.** The Organization and its Members, in pursuit of the Purposes stated in Article 1, shall act in accordance with the following Principles.

1. The Organization is based on the principle of the sovereign equality of all its Members.
2. All Members, in order to ensure to all of them the rights and benefits resulting from membership, shall fulfil in good faith the obligations assumed by them in accordance with the present Charter.
3. All Members shall settle their international disputes by peaceful means in such a manner that international peace and security, and justice, are not endangered.
4. All Members shall refrain in their international relations from the threat or use of force against the territorial integrity or political independence of any state, or in any other manner inconsistent with the Purposes of the United Nations.
5. All Members shall give the United Nations every assistance in any action it takes in accordance with the present Charter, and shall refrain from giving assistance to any state against which the United Nations is taking preventive or enforcement action.
6. The Organization shall ensure that states which are not Members of the United Nations act in accordance with these Principles so far as may be necessary for the maintenance of international peace and security.
7. Nothing contained in the present Charter shall authorize the United Nations to intervene it matters which are essentially within the domestic jurisdiction of any state or shall require the Members to submit such matters to settlement under the present Charter; but this principle shall not prejudice the application of enforcement measures under Chapter VII.

## II. MEMBERSHIP

**Art. 3.** The original Members of the United Nations shall be the states which, having participated in the United Nations Conference on International Organization at San Franciso, or having previously signed the Declaration by United Nations of 1 January 1942, sign the present Charter and ratify it in accordance with Article 110.

**Art. 4.** 1. Membership in the United Nations is open to all other peace-loving states which accept the obligations contained in the present Charter and, in the judgment of the Organization, are able and willing to carry out these obligations.

2. The admission of any such state to membership in the United Nations will be effected by a decision of the General Assembly upon the recommendation of the Security Council.

**Art. 5.** A Member of the United Nations against which preventive or enforcement action has been taken by the Security Council may be suspended from the exercise of the rights and privileges of membership by the General Assembly upon the recommendation of the Security Council. The exercise of these rights and privileges may be restored by the Security Council.

**Art 6.** A Member of the United Nations which has persistently violated the Principles contained in the present Charter may be expelled from the Organization by the General Assembly upon the recommendation of the Security Council.

## III. ORGANS

**Art. 7.** 1. There are established as the principal organs of the United Nations: a General Assembly, a Security Council, an Economic and Social Council, a Trusteeship Council, an International Court of Justice, and a Secretariat.

2. Such subsidiary organs as may be found necessary may be established in accordance with the present Charter.

**Art. 8.** The United Nations shall place no restrictions on the eligibility of men and women to participate in any capacity and under conditions of equality in its principal and subsidiary organs.

## IV. THE GENERAL ASSEMBLY

*Composition*

**Art. 9.** 1. The General Assembly shall consist of all the Members of the United Nations.

2. Each Member shall have not more than five representatives in the General Assembly.

*Functions and Powers*

**Art. 10.** The General Assembly may discuss any questions or any matters within the scope of the present Charter or relating to the powers and functions of any organs provided for in the present Charter, and, except as provided in Article 12, may make recommendations to the Members of the United Nations or to the Security Council or to both on any such questions or matters.

**Art. 11.** 1. The General Assembly may consider the general principles of co-operation in the maintenance of international peace and security, including the principles governing disarmament and the regulation of armaments, and may make recommendations with regard to such principles to the Members or to the Security Council or to both.

2. The General Assembly may discuss any questions relating to the maintenance of international peace and security brought before it by any Member of the United Nations, or by the Security Council, or by a state which is not a Member of the United Nations in accordance with Article 35, paragraph 2, and, except as provided in Article 12, may make recommendations with regard to any such question to the state or states concerned or to the Security Council or to both. Any such question on which action is necessary shall be referred to the Security Council by the General Assembly either before or after discussion.

3. The General Assembly may call the attention of the Security Council to situations which are likely to endanger international peace and security.

4. The powers of the General Assembly set forth in this Article shall not limit the general scope of Article 10.

**Art. 12.** 1. While the Security Council is exercising in respect of any dispute or situation the functions assigned to it in the present Charter, the General Assembly shall not make any recommendation with regard to that dispute or situation unless the Security Council so requests.

2. The Secretary-General, with the consent of the Security Council, shall notify the General Assembly at each session of any matters relative to the maintenance of international peace and security which are being dealt with by the Security Council and shall similarly notify the General Assembly, or the Members of the United Nations if the General Assembly is not in session, immediately the Security Council ceases to deal with such matters.

**Art. 13.** 1. The General Assembly shall initiate studies and make recommendations for the purpose of:

(*a*) promoting international co-operation in the political field and encouraging the progressive development of international law and its codification;

(*b*) promoting international co-operation in the economic, social, cultural, educational, and health fields, and assisting in the realization of human rights and fundamental freedoms for all without distinction as to race, sex, language, or religion.

2. The further responsibilities, functions and powers of the General Assembly with

respect to matters mentioned in paragraph 1 (*b*) above are set forth in Chapters IX and X.

**Art. 14.** Subject to the provisions of Article 12, the General Assembly may recommend measures for the peaceful adjustment of any situation, regardless of origin, which it deems likely to impair the general welfare or friendly relations among nations, including situations resulting from a violation of the provisions of the present Charter setting forth the Purpose and Principles of the United Nations.

**Art. 15.** 1. The General Assembly shall receive and consider annual and special reports from the Security Council; these reports shall include an account of the measures that the Security Council has decided upon or taken to maintain international peace and security.

2. The General Assembly shall receive and consider reports from the other organs of the United Nations.

**Art. 16.** The General Assembly shall perform such functions with respect to the international trusteeship system as are assigned to it under Chapters XII and XIII, including the approval of the trusteeship agreements for areas not designated as strategic.

**Art. 17.** 1. The General Assembly shall consider and approve the budget of the Organization.

2. The expenses of the Organization shall be borne by the Members as apportioned by the General Assembly.

3. The General Assembly shall consider and approve any financial and budgetary arrangements with specialized agencies referred to in Article 57 and shall examine the administrative budgets of such specialized agencies with a view to making recommendations to the agencies concerned.

*Voting*

**Art. 18.** 1. Each member of the General Assembly shall have one vote.

2. Decisions of the General Assembly on important questions shall be made by a two-thirds majority of the members present and voting. These questions shall include: recommendations with respect to the maintenance of international peace and security, the election of the non-permanent members of the Security Council, the election of the members of the Economic and Social Council, the election of members of the Trusteeship Council in accordance with paragraph 1 (*c*) of Article 86, the admission of new Members to the United Nations, the suspension of the rights and privileges of membership, the expulsion of Members, questions relating to the operation of the trusteeship system, and budgetary questions.

3. Decisions on other questions, including the determination of additional categories of questions to be decided by a two-thirds majority, shall be made by a majority of the members present and voting.

**Art. 19.** A Member of the United Nations which is in arrears in the payment of its financial contributions to the Organization shall have no vote in the General Assembly if the amount of its arrears equals or exceeds the amount of the contributions due from it for the preceding two full years. The General Assembly may, nevertheless, permit such a Member to vote if it is satisfied that the failure to pay is due to conditions beyond the control of the Member.

*Procedure*

**Art. 20.** The General Assembly shall meet in regular annual sessions and in such special sessions as occasion may require. Special sessions shall be convoked by the Secretary-General at the request of the Security Council or of a majority of the Members of the United Nations.

**Art. 21.** The General Assembly shall adopt its own rules of procedure. It shall elect its President for each session.

**Art. 22.** The General Assembly may establish such subsidiary organs as it deems necessary for the performance of its functions.

## V. THE SECURITY COUNCIL

*Composition*

**Art. 23.** 1. The Security Council shall consist of fifteen Members of the United Nations. The Republic of China, France, the Union of Soviet Socialist Republics, the United Kingdom of Great Britain and Northern Ireland, and the United States of America shall be permanent members of the Security Council. The General Assembly shall elect ten other Members of the United Nations to be non-permanent members of the Security Council, due regard being specially paid, in the first instance to the contribution of Members of the United Nations to the maintenance of international peace and security and to the other purposes of the Organization, and also to equitable geographical distribution.

2. The non-permanent members of the Security Council shall be elected for a term of two years. In the first election of the non-permanent members after the increase of the membership of the Security Council from eleven to fifteen, two of the four additional members shall be chosen for a term of one year. A retiring member shall not be elegible for immediate re-election.

3. Each member of the Security Council shall have one representative.

*Functions and Powers*

**Art. 24.** 1. In order to ensure prompt and effective action by the United Nations, its Members confer on the Security Council primary responsibility for the maintenance of international peace and security, and agree that in carrying out its duties under this responsibility the Security Council acts on their behalf.

2. In discharging these duties the Security Council shall act in accordance with the Purposes and Principles of the United Nations. The specific powers granted to the Security Council for the discharge of these duties are laid down in Chapters VI, VII, VIII, and XII.

3. The Security Council shall submit annual and, when necessary, special reports to the General Assembly for its consideration.

**Art. 25.** The Members of the United Nations agree to accept and carry out the decisions of the Security Council in accordance with the present Charter.

**Art. 26.** In order to promote the establishment and maintenance of international peace and security with the least diversion for armaments of the world's human and economic resources, the Security Council shall be responsible for formulating, with the assistance of the Military Staff Committee referred to in Article 47, plans to be submitted to the Members of the United Nations for the establishment of a system for the regulation of armaments.

*Voting*

**Art. 27.** 1. Each member of the Security Council shall have one vote.

2. Decisions of the Security Council on procedural matters shall be made by an affirmative vote of nine members.

3. Decisions of the Security Council on all other matters shall be made by an affirmative vote of nine members including the concurring votes of the permanent members; provided that, in decisions under Chapter VI, and under paragraph 3 of Article 52, a party to a dispute shall abstain from voting.

*Procedure*

**Art. 28.** 1. The Security Council shall be so organized as to be able to function continuously. Each member of the Security Council shall for this purpose be represented at all times at the seat of the Organization.

2. The Security Council shall hold periodic meetings at which each of its members may, if it so desires, be represented by a member of the government or by some other specially designated representative.

3. The Security Council may hold meetings at such places other than the seat of the Organization as in its judgment will best facilitate its work.

**Art. 29.** The Security Council may establish such subsidiary organs as it deems necessary for the performance of its functions.

**Art. 30.** The Security Council shall adopt its own rules of procedure, including the method of selecting its President.

**Art. 31.** Any Member of the United Nations which is not a member of the Security Council may participate, without vote, in the discussion of any question brought before the Security Council whenever the latter considers that the interests of that Member are specially affected.

**Art. 32.** Any Member of the United Nations which is not a member of the Security Council or any state which is not a Member of the United Nations, if it is a party to a dispute under consideration by the Security Council, shall be invited to participate, without vote, in the discussion relating to the dispute. The Security Council shall lay down such conditions as it deems just for the participation of a state which is not a Member of the United Nations.

## VI. PACIFIC SETTLEMENT OF DISPUTES

**Art. 33.** 1. The parties to any dispute, the continuance of which is likely to endanger the maintenance of international peace and security, shall, first of all, seek a solution by negotiation, enquiry, mediation, conciliation, arbitration, judical settlement, resort to regional agencies or arrangements, or other peaceful means of their own choice.

2. The Security Council shall, when it deems necessary, call upon the parties to settle their dispute by such means.

**Art. 34.** The Security Council may investigate any dispute, or any situation which might lead to international friction or give rise to a dispute, in order to determine whether the continuance of the dispute or situation is likely to endanger the maintenance of international peace and security.

**Art. 35.** 1. Any Member of the United Nations may bring any dispute, or any situation of the nature referred to in Article 34, to the attention of the Security Council or of the General Assembly.

2. A state which is not a Member of the United Nations may bring to the attention of the Security Council or of the General Assembly any dispute to which it is a party if it accepts in advance, for the purposes of the dispute, the obligations of pacific settlement provided in the present Charter.

3. The proceedings of the General Assembly in respect of matters brought to its attention under this Article will be subject to the provisions of Articles 11 and 12.

**Art. 36.** 1. The Security Council may, at any stage of a dispute of the nature referred to in Article 33 or of a situation of like nature, recommend appropriate procedures or methods of adjustment.

2. The Security Council should take into consideration any procedures for the settlement of the dispute which have already been adopted by the parties.

3. In making recommendations under this Article the Security Council should also take into consideration that legal disputes should as a general rule be referred by the parties to the International Court of Justice in accordance with the provisions of the Statute of the Court.

**Art. 37.** 1. Should the parties to a dispute of the nature referred to in Article 33 fail to settle it by the means indicated in that Article, they shall refer it to the Security Council.

2. If the Security Council deems that the continuance of the dispute is in fact likely to endanger the maintenance of international peace and security, it shall decide whether to take action under Article 36 or to recommend such terms of settlement as it may consider appropriate.

**Art. 38.** Without prejudice to the provisions of Articles 33 to 37, the Security Council may, if all the parties to any dispute so request, make recommendations to the parties with a view to a pacific settlement of the dispute.

## VII. ACTION WITH RESPECT TO THREATS TO THE PEACE, BREACHES OF THE PEACE, AND ACTS OF AGGRESSION

**Art. 39.** The Security Council shall determine the existence of any threat to the peace, breach of the peace, or act of aggression and shall make recommendations, or decide what measures shall be taken in accordance with Articles 41 and 42, to maintain or restore international peace and security.

**Art. 40.** In order to prevent an aggravation of the situation, the Security Council may, before making the recommendations or deciding upon the measures provided for in Article 39, call upon the parties concerned to comply with such provisional measures as it deems necessary or desirable. Such provisional measures shall be without prejudice to the rights, claims, or position of the parties concerned. The Security Council shall duly take account of failure to comply with such provisional measures.

**Art. 41.** The Security Council may decide what measures not involving the use of armed forces are to be employed to give effect to its decisions, and it may call upon the Members of the United Nations to apply such measures. These may include complete or partial interruption of economic relations and of rail, sea, air, postal, telegraphic, radio, and other means of communication, and the severance of diplomatic relations.

**Art. 42.** Should the Security Council consider that measures provided for in Article 41 would be inadequate or have proved to be inadequate, it may take such action by air, sea, or land forces as may be necessary to maintain or restore international peace and security. Such action may include demonstrations, blockade, and other operations by air, sea, or land forces of Members of the United Nations.

**Art. 43.** 1. All Members of the United Nations, in order to contribute to the maintenance of international peace and security, undertake to make available to the Security Council, on its call and inaccordance with a special agreement or agreements, armed forces, assistance, and facilities, including rights of passage, necessary for the purpose of maintaining international peace and security. 2. Such agreement or agreements shall govern the numbers and types of forces, their degree of readiness and general location, and the nature of the facilities and assistance to be provided.

3. The agreement or agreements shall be negotiated as soon as possible on the initiative of the Security Council. They shall be concluded between the Security Council and Members or between the Security Council and groups of Members and shall be subject to ratification by the signatory states in accordance with their respective constitutional processes.

**Art. 44.** When the Security Council has decided to use force it shall, before calling upon a Member not represented on it to provide armed forces in fulfilment of the obligations assumed under Article 43, invite that Member, if the Member so desires, to participate in the decisions of the Security Council concerning the employment of contingents of that Member's armed forces.

**Art. 45.** In order to enable the United Nations to take urgent military measures, Members shall hold immediately available national air-force contingents for combined international enforcement action. The strength and degree of readiness of these contingents and plans for their combined action shall be determined, within the limits laid down in the special agreement or agreements referred to in Article 43, by the Security Council with the assistance of the Military Staff Committee.

**Art. 46.** Plans for the application of armed force shall be made by the Security Council with the assistance of the Military Staff Committee.

**Art. 47.** 1. There shall be established a Military Staff Committee to advise and assist the Security Council on all questions relating to the Security Council's military requirements for the maintenance of international peace and security, the employment and command of forces placed at its disposal, the regulation of armaments, and possible disarmament.

2. The Military Staff Committee shall consist of the Chiefs of Staff of the permanent members of the Security Council or their representatives. Any Member of the United Nations not permanently represented on the Committee shall be invited by the Committee to be associated with it when the efficient discharge of the Committee's responsibilities requires the participation of that Member in its work.

3. The Military Staff Committee shall be responsible under the Security Council for the strategic direction of any armed forces placed at the disposal of the Security Council. Questions relating to the command of such forces shall be worked out subsequently.

4. The Military Staff Committee, with the authorization of the Security Council and after consultation with appropriate regional agencies, may establish regional subcommittees.

**Art. 48.** 1. The action required to carry out the decisions of the Security Council for the maintenance of international peace and security shall be taken by all the Members of the United Nations or by some of them, as the Security Council may determine.

2. Such decisions shall be carried out by the Members of the United Nations directly and through their action in the appropriate international agencies of which they are members.

**Art. 49.** The Members of the United Nations shall join in affording mutual assistance in carrying out the measures decided upon by the Security Council.

**Art. 50.** If preventive or enforcement measures against any state are taken by the Security Council any other state, whether a Member of the United Nations or not, which finds itself confronted with special economic problems arising from the carrying out of those measures shall have the right to consult the Security Council with regard to a solution of those problems.

**Art. 51.** Nothing in the present Charter shall impair the inherent right of individual or collective self-defense if an armed attack occurs against a Member of the United Nations, until the Security Council has taken measures necessary to maintain international peace and security. Measures taken by Members in the exercise of this right of self-defense shall be immediately reported to the Security Council and shall not in any way affect the authority and responsibility of the Security Council under the present Charter to take at any time such action as it deems necessary in order to maintain or restore international peace and security.

## VIII. REGIONAL ARRANGEMENTS

**Art. 52.** 1. Nothing in the present Charter precludes the existence of regional arrangements or agencies for dealing with such matters relating to the maintenance of international peace and security as are appropriate for regional action, provided that such arrangements or agencies and their activities are consistent with the Purposes and Principles of the United Nations.

2. The Members of the United Nations entering into such arrangements or constituting such agencies shall make every effort to achieve pacific settlement of local disputes through such regional arrangements or by such regional agencies before referring them to the Security Council.

3. The Security Council shall encourage the development of pacific settlement of local disputes through such regional arrangements or by such regional agencies either on the initiative of the states concerned or by reference from the Security Council.

4. This Article in no way impairs the application of Articles 34 and 35.

**Art. 53.** The Security Council shall, where appropriate, utilize such regional arrangements or agencies for enforcement action under its authority. But no enforcement action shall be taken under regional arrangements or by regional agencies without the authorization of the Security Council, with the exception of measures against any enemy state, as defined in paragraph 2 of this Article, provided for pursuant to Article 107 or in regional arrangements directed against renewal of aggressive policy on the part of any such state, until such time as the Organization may, on request of the Governments concerned, be charged with the responsibility for preventing further aggression by such a state.

2. The term enemy state as used in paragraph 1 of this Article applies to any state which during the Second World War has been an enemy of any signatory of the present Charter.

**Art. 54.** The Security Council shall at all times be kept fully informed of activities undertaken or in contemplation under regional arrangements or by regional agencies for the maintenance of international peace and security.

## IX. INTERNATIONAL ECONOMIC AND SOCIAL CO-OPERATION

**Art. 55.** With a view to the creation of conditions of stability and well-being which are necessary for peaceful and friendly relations among nations based on respect for the principle of equal rights and self-determination of peoples, the United Nations shall promote:

(*a*) higher standards of living, full employment, and conditions of economic and social progress and development;

(*b*) solutions of international economic, social, health, and related problems; and international cultural and educational co-operation; and

(*c*) universal respect for, and observance of, human rights and fundamental freedoms for all without distinction as to race, sex, language, or religion.

**Art. 56.** All Members pledge themselves to take joint and separate action in co-operation with the Organization for the achievement of the purposes set forth in Article 55.

**Art. 57.** 1. The various specialized agencies, established by inter-governmental agreement and having wide international responsibilities, as defined in their basic instruments, in economic, social, cultural, educational, health, and related fields, shall be brought into relationship with the United Nations in accordance with the provisions of Article 63.

2. Such agencies thus brought into relationship with the United Nations are hereinafter referred to as specialized agencies.

**Art. 58.** The Organization shall make recommendations for the co-ordination of the policies and activities of the specialized agencies.

**Art. 59.** The Organization shall, where appropriate, initiate negotiations among the states concerned for the creation of any new specialized agencies required for the accomplishment of the purposes set forth in Article 55.

**Art. 60.** Responsibility for the discharge of the functions of the Organization set forth in this Chapter shall be vested in the General Assembly and, under the authority of the General Assembly, in the Economic and Social Council, which shall have for this purpose the powers set forth in Chapter X.

## X. THE ECONOMIC AND SOCIAL COUNCIL

### *Composition*

**Art. 61.** 1. The Economic and Social Council shall consist of twenty-seven Members of the United Nations elected by the General Assembly.

2. Subject to the provisions of paragraph 3, nine members of the Economic and Social Council shall be elected each year for a term of three years. A retiring member shall be eligible for immediate re-election.

3. At the first election after the increase in the membership of the Economic and Social Council from eighteen to twenty-seven members, in addition to the members elected in place of the six members whose term of office expires at the end of that year, nine additional members shall be elected. Of these nine additional members, the term of office of three members so elected shall expire at the end of one year, and of three other members at the end of two years, in accordance with arrangements made by the General Assembly.

4. Each member of the Economic and Social Council shall have one representative.

### *Functions and Powers*

**Art. 62.** 1. The Economic and Social Council may make or initiate studies and reports with respect to international economic, social, cultural, educational, health, and related matters and may make recommendations with respect to any such matters to the General Assembly, to the Members of the United Nations, and to the specialized agencies concerned.

2. It may make recommendations for the purpose of promoting respect for, and observance of, human rights and fundamental freedoms for all.

3. It may prepare draft conventions for submission to the General Assembly, with respect to matters falling within its competence.

4. It may call, in accordance with the rules prescribed by the United Nations, international conferences on matters falling within its competence.

**Art. 63.** 1. The Economic and Social Council may enter into agreements with any of the agencies referred to in Article 57, defining the terms on which the agency concerned shall be brought into relationship with the United Nations. Such agreements shall be subject to approval by the General Assembly.

2. It may co-ordinate the activities of the specialized agencies through consultation with and recommendations to such agencies and through recommendations to the General Assembly and to the Members of the United Nations.

**Art. 64.** 1. The Economic and Social Council may take appropriate steps to obtain regular reports from the specialized agencies. It may make arrangements with the Members of the United Nations and with the specialized agencies to obtain reports on the steps taken to give effect to its own recommendations and to recommendations on matters falling within its competence made by the General Assembly.

2. It may communicate its observations on these reports to the General Assembly.

**Art. 65.** The Economic and Social Council may furnish information to the Security Council and shall assist the Security Council upon its request.

**Art. 66.** 1. The Economic and Social Council shall perform such functions as fall within its competance in connexion with the carrying out of the recommendations of the General Assembly.

2. It may, with the approval of the General Assembly, perform services at the request of Members of the United Nations and at the request of specialized agencies.

3. It shall perform such other functions as are specified elsewhere in the present Charter or as may be assigned to it by the General Assembly.

*Voting*

**Art. 67.** 1. Each member of the Economic and Social Council shall have one vote.

2. Decisions of the Economic and Social Council shall be made by a majority of the members present and voting.

*Procedure*

**Art. 68.** The Economic and Social Council shall set up commissions in economic and social fields and for the promotion of human rights, and such other commissions as may be required for the performance of its functions.

**Art. 69.** The Economic and Social Council shall invite any Member of the United Nations to participate, without vote, in its deliberations on any matter of particular concern to that Member.

**Art. 70.** The Economic and Social Council may make arrangements for representatives of the specialized agencies to participate, without vote, in its deliberations and in those of the commissions established by it, and for its representatives to participate in the deliberations of the specialized agencies.

**Art. 71.** The Economic and Social Council may make suitable arrangements for consultation with non-governmental organizations which are concerned with matters within its competence. Such arrangements may be made with international organizations and, where appropriate, with national organizations after consultation with the Member of the United Nations concerned.

**Art. 72.** 1. The Economic and Social Council shall adopt its own rules of procedure, including the method of selecting its President.

2. The Economic and Social Council shall meet as required in accordance with its rules, which shall include provision for the convening of meetings on the request of a majority of its members.

## XI. DECLARATION REGARDING NON-SELF-GOVERNING TERRITORIES

**Art. 73.** Members of the United Nations which have or assume responsibilities for the administration of territories whose peoples have not yet attained a full measure of self-government recognize the principle that the interests of the inhabitants of these territories are paramount, and accept as a sacred trust the obligation to promote to the utmost, within the system of international peace and security established by the present Charter, the well-being of the inhabitants of these territories, and, to this end:

(*a*) to ensure, with due respect for the culture of the peoples concerned, their political economic, social, and educational advancement, their just treatment, and their protection against abuses;

(*b*) to develop self-government, to take due account of the political aspirations of the peoples, and to assist them in the progressive development of their free political institutions, according to the particular circumstances of each territory and its peoples and their varying stages of advancement;

(*c*) to further international peace and security;

(*d*) to promote constructive measures of development, to encourage research, and to co-operate with one another and, when and where appropriate, with specialized international bodies with a view to the practical achievement of the social economic, and scientific purposes set forth in this Article; and

(*e*) to transmit regularly to the Secretary-General for information purposes, subject to such limitation as security and constitutional considerations may require, statistical and other information of a technical nature relating to economic, social, and educational conditions in the territories for which they are respectively responsible other than those territories to which Chapters XII and XIII apply.

**Art. 74.** Members of the United Nations also agree that their policy in respect of the territories to which this Chapter applies, no less than in respect of their metropolitan areas, must be based on the general principle of goodneighbourliness, due account being taken of the interests and well-being of the rest of the world, in social, economic, and commercial matters.

## XII. INTERNATIONAL TRUSTEESHIP SYSTEM

**Art. 75.** The United Nations shall establish under its authority an international trusteeship system for the administration and supervision of such territories as may be placed thereunder by subsequent individual agreements. These territories are hereinafter referred to as trust territories.

**Art. 76.** The basic objectives of the trusteeship system, in accordance with the Purposes of the United Nations laid down in Article 1 of the present Charter, shall be:

(*a*) to further international peace and security;

(*b*) to promote the political, economic, social, and educational advancement of the inhabitants of the trust territories, and their progressive development towards self-government or independence as may be appropriate to the particular circumstances of each territory and its peoples and the freely expressed wishes of the peoples concerned, and as may be provided by the terms of each trusteeship agreement;

(*c*) to encourage respect for human rights and for fundamental freedoms for all without distinction as to race, sex, language, or religion, and to encourage recognition of the interdependence of the peoples of the world; and

(*d*) to ensure equal treatment in social, economic, and commercial matters for all Members of the United Nations and their nationals, and also equal treatment for the latter in the administration of justice, without prejudice to the attainment of the foregoing objectives and subject to the provisions of Article 80.

**Art. 77.** 1. The trusteeship system shall apply to such territories in the following categories as may be placed thereunder by means of trusteeship agreements:

(*a*) territories now held under mandate;

(*b*) territories which may be detached from enemy states as a result of the Second World War; and

(*c*) territories voluntarily placed under the system by states responsible for their administration.

2. It will be a matter for subsequent agreement as to which territories in the foregoing categories will be brought under the trusteeship system and upon what terms.

**Art. 78.** The trusteeship system shall not apply to territories which have become Members of the United Nations, relationship among which shall be based on respect for the principle of sovereign equality.

**Art. 79.** The terms of trusteeship for each territory to be placed under the trusteeship system, including any alteration or amendment, shall be agreed upon by the states directly concerned, including the mandatory power in the case of territories held under mandate by a Member of the United Nations, and shall be approved as provided for in Articles 83 and 85.

**Art. 80.** 1. Except as may be agreed upon in individual trusteeship agreements, made under Articles 77, 79, and 81, placing each territory under the trusteeship system, and until such agreements have been concluded, nothing in this Chapter shall be construed in or of itself to alter in any manner the rights whatsoever of any states or any peoples or the terms of existing international instruments to which Members of the United Nations may respectively be parties.

2. Paragraph 1 of this Article shall not be interpreted as giving grounds for delay or postponement of the negotiation and conclusion of agreements for placing mandated and other territories under the trusteeship system as provided for in Article 77.

**Art. 81.** The trusteeship agreement shall in each case include the terms under which the trust territory will be administered and designate the authority which will exercise the administration of the trust territory. Such authority, hereinafter called the administering authority, may be one or more states or the Organization itself.

**Art. 82.** There may be designated, in any trusteeship agreement, a strategic area or areas which may include part of all of the trust territory to which the agreement applies, without prejudice to any special agreement or agreements made under Article 43.

**Art. 83.** All functions of the United Nations relating to strategic areas, including the approval of the terms of the trusteeship agreements and of their alteration of amendment, shall be exercised by the Security Council.

2. The basic objectives set forth in Article 76 shall be applicable to the people of each strategic area.

3. The Security Council shall, subject to the provisions of the trusteeship agreements and without prejudice to security considerations, avial itself of the assistance of the Trusteeship Council to perform those functions of the United Nations under the trusteeship system relating to political, economic, social, and educational matters in the strategic areas.

**Art. 84.** It shall be the duty of the administering authority to ensure that the trust territory shall play its part in the maintenance of international peace and security. To this end the administering authority may make use of volunteer forces, facilities, and assistance from the trust territory in carrying out the obligations towards the Security Council undertaken in this regard by the administering authority, as well as for local defense and the maintenance of law and order within the trust territory.

**Art. 85.** 1. The functions of the United Nations with regard to trusteeship agreements for all areas not designated at strategic, including the approval of the terms of the trusteeship agreements and of their alteration or amendment, shall be exercised by the General Assembly.

2. The Trusteeship Council, operating under the authority of the General Assembly, shall assist the General Assembly in carrying out these functions.

## XIII. THE TRUSTEESHIP COUNCIL

### *Composition*

**Art. 86.** 1. The Trusteeship Council shall consist of the following Members of the United Nations:

(*a*) those Members administering trust territories;

(*b*) such of those Members mentioned by name in Article 23 as are not administering trust territories; and (*c*) as many other Members elected for three-year terms by the General Assembly as may be necessary to ensure that the total number of members of the Trusteeship Council is equally divided between those Members of the United Nations which administer trust territories and those which do not.

2. Each member of the Trusteeship Council shall designate one specially qualified person te represent it therein.

*Functions and Powers*

**Art. 87.** The General Assembly and, under its authority, the Trusteeship Council, in carrying out their functions, may:

(*a*) consider reports submitted by the administering authority;

(*b*) accept petitions and examine them in consultation with the administering authority;

(*c*) provide for periodic visits to the respective trust territories at times agreed upon with the administering authority; and

(*d*) take these and other actions in conformity with the terms of the trusteeship agreements.

**Art. 88.** The trusteeship Council shall formulate a questionnaire on the political, economic, social, and educational advancement of the inhabitants of each trust territory, and the administering authority for each trust territory within the competence of the General Assembly shall make an annual report to the General Assembly upon the basis of such questionnaire.

*Voting*

**Art. 89.** 1. Each member of the Trusteeship Council shall have one vote.

2. Decisions of the Trusteeship Council shall be made by a majority of the members present and voting.

*Procedure*

**Art. 90.** 1. The Trusteeship Council shall adopt its own rules of procedure, including the method of selecting its President.

2. The Trusteeship Council shall meet as required in accordance with its rules, which shall include provision for the convening of meetings on the request of a majority of its members.

**Art. 91.** The Trusteeship Council shall, when appropriate, avail itself of the assistance of the Economic and Social Council and of the specialized agencies in regard to matters with which they are respectively concerned.

## XIV. THE INTERNATIONAL COURT OF JUSTICE

**Art. 92.** The International Court of Justice shall be the Principal judicial organ of the United Nations. It shall function in accordance with the annexed Statute, which is based upon the Statute of the Permanent Court of International Justice and forms an integral part of the present Charter.

**Art. 93.** 1. All Members of the United Nations are *ipso facto* parties to the Statute of the International Court of Justice.

2. A state which is not a Member of the United Nations may become a party to the Statute of the International Court of Justice on conditions to be determined in each case by the General Assembly upon the recommendation of the Security Council.

**Art. 94.** 1. Each Member of the United Nations undertakes to comply with the decision of the International Court of Justice in any case to which it is a party.

2. If any party to a case fails to perform the obligations incumbent upon it under a judgment rendered by the Court, the other party may have recourse to the Security Council, which may, if it deems necessary, make recommendations or decide upon measures to be taken to give effect to the judgment.

**Art. 95.** Nothing in the present Charter shall prevent Members of the United Nations from entrusting the solution of their differences to other tribunals by virtue of agreements already in existence or which may be concluded in the future.

**Art. 96.** 1. The General Assembly or the Security Council may request the International Court of Justice to give an advisory opinion on any legal question.

2. Other organs of the United Nations and specialized agencies, which may at any time be so authorized by the General Assembly, may also request advisory opinions of the Court on legal questions arising within the scope of their activities.

## XV. THE SECRETARIAT

**Art. 97.** The Secretariat shall comprise a Secretary-General and such staff as the Organization may require. The Secretary-General shall be appointed by the General Assembly upon the recommendation of the Security Council. He shall be the chief administrative officer of the Organization.

**Art. 98.** The Secretary-General shall act in that capacity in all meetings of the General Assembly, of the Security Council, of the Economic and Social Council, and of the Trusteeship Council, and shall perform such other functions as are entrusted to him by these organs. The Secretary-General shall make an annual report to the General Assembly on the work of the Organization.

**Art. 99.** The Secretary-General may bring to the attention of the Security Council any matter which in his opinion may threaten the maintenance of international peace and security.

**Art. 100.** 1. In the performance of their duties the Secretary-General and the staff shall not seek or receive instructions from any government or from any other authority external to the Organization. They shall refrain from any action which might reflect on their position as international officials responsible only to the Organization.

2. Each Member of the United Nations undertakes to respect the exclusively international character of the responsibilities of the Secretary-General and the staff and not to seek to influence them in the discharge of their responsibilities.

**Art. 101.** 1. The staff shall be appointed by the Secretary-General under regulations established by the General Assembly.

2. Appropriate staffs shall be permanently assigned to the Economic and Social Council, the Trusteeship Council, and, as required, to other organs of the United Nations. These staffs shall form a part of the Secretariat.

3. The paramount consideration in the employment of the staff and in the determination of the conditions of service shall be the necessity of securing the highest standards of efficiency, competence, and integrity. Due regard shall be paid to the importance of recruiting the staff on as wide a geographical basis as possible.

## XVI. MISCELLANEOUS PROVISIONS

**Art. 102.** 1. Every treaty and every international agreement entered into by any Member of the United Nations after the present Charter comes into force shall as soon as possible be registered with the Secretariat and published by it.

2. No party to any such treaty or international agreement which has not been registered in accordance with the provisions of paragraph 1 of this Article may invoke that treaty or agreement before any organ of the United Nations.

**Art. 103.** In the event of a conflict between the obligations of the Members of the United Nations under the present Charter and their obligations under any other international agreement, their obligations under the present Charter shall prevail.

**Art. 104.** The Organization shall enjoy in the territory of each of its Members such legal capacity as may be necessary for the exercise of its functions and the fulfilment of its purposes.

**Art. 105.** 1. The Organization shall enjoy in the territory of each of its Members such privileges and immunities as are necessary for the fulfilment of its purposes.

2. Representatives of the Members of the United Nations and officials of the Organization shall similarly enjoy such privileges and immunities as are necessary for the independent exercise of their functions in connexion with the Organization.

3. The General Assembly may make recommendations with a view to determining

the details of the application of paragraphs 1 and 2 of this Article or may propose convention to the Members of the United Nations for this purpose.

## XVII. TRANSITIONAL SECURITY ARRANGEMENTS

**Art. 106.** Pending the coming into force of such special agreements referred to in Article 43 as in the opinion of the Security Council enable it to begin the exercise of its responsibilities under Article 42, the parties to the Four-Nation Declaration, signed at Moscow, October 30, 1943, and France, shall, in accordance with the provisions of paragraph 5 of that Declaration, consult with one another and as occasion requires with other Members of the United Nations with a view to such joint action on behalf of the Organization as may be necessary for the purpose of maintaining international peace and security.

**Art. 107.** Nothing in the present Charter shall invalidate or preclude action, in relation to any state which during the Second World War has been an enemy of any signatory to the present Charter, taken or authorized as a result of that war by the Governments having responsibility for such action.

## XVIII. AMENDMENTS

**Art. 108.** Amendments to the present Charter shall come into force for all Members of the United Nations when they have been adopted by a vote of two thirds of the members of the General Assembly and ratified in accordance with their respective constitutional processes by two thirds of the Members of the United Nations, including all the permanent members of the Security Council.

**Art. 109.** 1. A General Conference of the Members of the United Nations for the purpose of reviewing the present Charter may be held at a date and place to be fixed by a two-thirds vote of the Security Council. Each Member of the United Nations shall have one vote in the conference.

2. Any alteration of the present Charter recommended by a two thirds vote of the conference shall take effect when ratified in accordance with their respective constitutional processes by two thirds of the Members of the United Nations including all the permanent members of the Security Council.

3 If such conference has not been held before the tenth annual session of the General Assembly following the coming into force of the present Charter, the proposal to call such a conference shall be placed on the agenda of that session of the General Assembly, and the conference shall be held if so decided by a majority vote of the members of the General Assembly and by a vote of any seven members of the Security Council.

## XIX. RATIFICATION AND SIGNATURE

**Art. 110.** 1. The present Charter shall be ratified by the signatory states in accordance with their respective constitutional processes.

2. The ratifications shall be deposited with the Government of the United States of America, which shall notify all the signatory states of each deposit as well as the Secretary-General of the Organizations when he has been appointed.

3. The present Charter shall come into force upon the deposit of ratification by the Republic of China, France, the Union of Soviet Socialist Republics, the United Kingdom of Great Britain and Northern Ireland, and the United States of America, and by a majority of the other signatory states. A protocol of the ratifications deposited shall thereupon be drawn up by the Governments of the United States of America which shall communicate copies thereof to all the signatory states.

4. The states signatory to the present Charter which ratify it after it has come into force will become original members of the United Nations on the date of the deposit of their respective ratifications.

**Art. 111.** The Present Charter, of which the Chinese, French, Russian, English, and Spanish texts are equally authentic, shall remain deposited in the archives of the Government of the United States of America. Duly certified copies thereof shall be transmitted by that Government to the Governments of the other signatory states.

*In faith whereof* the representatives of the Governments of the United Nations have signed the present Charter.

*Done* at the city of San Francisco the twenty-sixth day of June, one thousand nine hundred and forty-five.